T0304509

Multiple Criteria Decision Making Applications in Environmentally Conscious Manufacturing and Product Recovery

Multiple Criteria Decision Making Applications in Environmentally Conscious Manufacturing and Product Recovery

By

Surendra M. Gupta and Mehmet Ali Ilgin

CRC Press
Taylor & Francis Group
Boca Raton London New York

CRC Press is an imprint of the
Taylor & Francis Group, an **informa** business

CRC Press
Taylor & Francis Group
6000 Broken Sound Parkway NW, Suite 300
Boca Raton, FL 33487-2742

© 2018 by Taylor & Francis Group, LLC

CRC Press is an imprint of Taylor & Francis Group, an Informa business

No claim to original U.S. Government works

Printed on acid-free paper

International Standard Book Number-13: 978-1-4987-0065-8 (Hardback)

Visit the Taylor & Francis Web site at
http://www.taylorandfrancis.com

and the CRC Press Web site at
http://www.crcpress.com

To Sharda Gupta, Monica Gupta, and Neil Gupta
             – **Surendra M. Gupta**

Bahar Sarı Ilgın
             – **Mehmet Ali Ilgin**

Contents

Preface

Shorter product life cycles and premature disposal of products are two major consequences of rapid technological advancement in product technology. This trend has resulted in the dramatic decrease of natural resources and an alarming decrease in the number of landfill sites. As a remedy for these problems, local governments are imposing stricter environmental regulations. To comply with these regulations and have a better environmental image in society, firms invest in environmentally conscious manufacturing, which involves the development of manufacturing methods that comply with environmental legislation and requirements considering all phases in a product's life cycle (i.e., from conceptual design to end-of-life [EOL] processing). They also set up specific facilities for product recovery, which can be defined as the minimization of the amount of waste sent to landfills by recovering materials and components from returned or EOL products via disassembly, recycling, and remanufacturing.

To solve the problems associated with environmentally conscious manufacturing and product recovery (ECMPRO), researchers have developed various algorithms, models, heuristics, and software. Among them, multiple criteria decision making (MCDM) techniques have received considerable attention from researchers, as those techniques can simultaneously consider more than one objective. Moreover, they are very good at modeling conflicting objectives, a common characteristic of ECMPRO issues (e.g., maximization of revenue from product recovery operations vs. minimization of environmental consequences of those operations).

In this book, we demonstrate how MCDM techniques may facilitate solutions to the problems associated with ECMPRO. The MCDM techniques considered include goal programming, linear physical programming, data envelopment analysis, analytical hierarchy process, analytical network process, DEMATEL, TOPSIS, ELECTRE, PROMETHEE, VIKOR, MACBETH and gray relational analysis.

The examples considered in this book can serve as starting points for researchers to build bodies of knowledge in the fast and growing areas of ECMPRO. Moreover, practitioners can benefit from the book by understanding the steps to be followed while applying a particular MCDM technique to ECMPRO issues.

Acknowledgments

We want to thank Taylor & Francis Group for their commitment to publish and stimulate innovative ideas. We also express our sincere appreciation to Cindy Renee Carelli and the Taylor & Francis Group staff for providing unwavering support at every stage of this project.

Finally, we acknowledge our families, to whom this book is lovingly dedicated, for their unconditional love and support for making this book a reality.

Surendra M. Gupta, PhD
Boston, Massachusetts

Mehmet Ali Ilgin, PhD
Manisa, Turkey

Authors

Surendra M. Gupta is a professor of mechanical and industrial engineering and the director of the Laboratory for Responsible Manufacturing, Northeastern University. He received his BE in electronics engineering from Birla Institute of Technology and Science, MBA from Bryant University, and MSIE and PhD in industrial engineering from Purdue University. He is a registered professional engineer in the state of Massachusetts. Dr. Gupta's research interests span the areas of production/manufacturing systems and operations research. He is mostly interested in environmentally conscious manufacturing, reverse and closed-loop supply chains, disassembly modeling, and remanufacturing. He has authored or coauthored ten books and well over 500 technical papers published in edited books, journals, and international conference proceedings. His publications have received over 10,000 citations from researchers all over the world in journals, proceedings, books, and dissertations. He has traveled to all seven continents, namely, Africa, Antarctica, Asia, Australia, Europe, North America, and South America, and presented his work at international conferences on six continents. Dr. Gupta has taught over 150 courses in such areas as operations research, inventory theory, queuing theory, engineering economy, supply-chain management, and production planning and control. Among the many recognitions received, he is the recipient of the outstanding research award and the outstanding industrial engineering professor award (in recognition of teaching excellence) from Northeastern University, as well as a national outstanding doctoral dissertation advisor award.

Mehmet Ali Ilgin is an assistant professor of industrial engineering at Manisa Celal Bayar University. He holds a PhD in industrial engineering from Northeastern University in Boston, and BS and MS in industrial engineering from Dokuz Eylul University. His research interests are in the areas of environmentally conscious manufacturing, product recovery, remanufacturing, reverse logistics, spare parts inventory management, and simulation. He has published a number of research papers in refereed international journals such as *Computers and Industrial Engineering*, *Robotics and Computer Integrated Manufacturing*, and the *International Journal of Advanced Manufacturing Technology*. He is a coauthor of the CRC Press book *Remanufacturing Modeling and Analysis*.

1

Multiple Criteria Decision Making in Environmentally Conscious Manufacturing and Product Recovery

1.1 Introduction

Shorter product life cycles and premature disposal of products are two major consequences of rapid technological advancement in product technology. This trend has resulted in the dramatic decrease of natural resources and an alarming decrease in the number of landfill sites. As a remedy for these problems, local governments are imposing stricter environmental regulations. To comply with these regulations and have a better environmental image in society, firms invest in environmentally conscious manufacturing, which involves the development of manufacturing methods that comply with environmental legislation and requirements, considering all phases in a product's life cycle (i.e., from conceptual design to end-of-life [EOL] processing). They also set up specific facilities for product recovery, which can be defined as the minimization of the amount of waste sent to landfills by recovering materials and components from returned or EOL products via disassembly, recycling, and remanufacturing.

To solve the problems associated with environmentally conscious manufacturing and product recovery (ECMPRO), researchers have developed various algorithms, models, heuristics, and software (Zhang et al. 1997; Gungor and Gupta 1999; Ilgin and Gupta 2010). Among them, multiple criteria decision making (MCDM) techniques have received considerable attention from researchers, since those techniques can simultaneously consider more than one objective (Toloie-Eshlaghy and Homayonfar 2011). Moreover, they are very good at modeling conflicting objectives, a common characteristic of ECMPRO issues (e.g., maximization of revenue from product recovery operations vs. minimization of environmental consequences of those operations).

This chapter presents an overview of the state of the art of MCDM techniques applied in ECMPRO problems. The reviewed studies are categorized into five main areas, namely, quantitative techniques, qualitative techniques, mixed techniques, heuristics and metaheuristics, and simulation (see Figure 1.1). Studies are classified into subcategories (where appropriate) within each main area. In Section 1.2, the studies employing quantitative MCDM techniques are discussed. Section 1.3 presents the studies using qualitative techniques. The studies integrating two or more qualitative and/or quantitative techniques are reviewed in Section 1.4. Section 1.5 provides a detailed analysis of heuristics- and metaheuristics-based studies. An overview of simulation-based studies is presented in Section 1.6. Finally, Section 1.7 provides some concluding remarks and future research directions.

1.2 Quantitative Techniques

These techniques aim at achieving the optimal or aspired goals by considering the various interactions within the given constraints (Tzeng and Huang 2011). They are also known as multiobjective decision making techniques. In this section, we categorize the ECMPRO applications of quantitative techniques into five segments; namely, goal programming, fuzzy goal programming, physical programming, data envelopment analysis, and other mathematical models.

1.2.1 Goal Programming

Goal programming is an extension of linear programming, due to its ability to handle multiple and often conflicting objectives (Ignizio 1976). Two variants of goal programming are prevalent in the literature. The first one is known as lexicographic or preemptive goal programming, while the second one is termed weighted or nonpreemptive goal programming. Preemptive goal programming assumes that all goals can be clearly prioritized and that satisfying a higher-priority goal should carry more importance than satisfying a lower-priority goal. Nonpreemptive goal programming assumes that all goals should be pursued. However, in this case, all deviations from the goals are multiplied by some weights (based on their relative importance) and summed up to form a single utility function that is optimized.

Kongar and Gupta (2000) present a preemptive integer goal-programming model for the disassembly-to-order (DTO) process, so as to satisfy various economic, physical, and environmental goals simultaneously. Imtanavanich and Gupta (2006a) use preemptive goal programming to solve the multicriteria DTO problem under stochastic yields. Massoud and Gupta (2010b) extend Imtanavanich and Gupta (2006a) by using preemptive goal programming to solve a similar problem under stochastic yields, limited supply, and quantity discount.

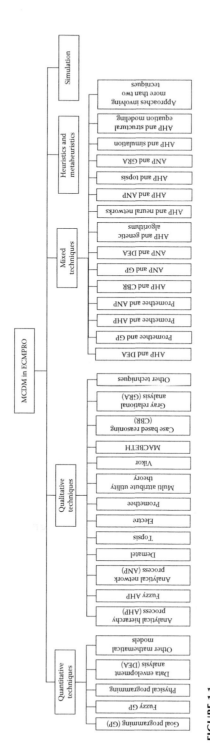

FIGURE 1.1

Classification of MCDM techniques used to solve ECMPRO issues.

McGovern and Gupta (2008) use lexicographic goal programming to solve the disassembly line balancing problem, which involves the determination of a sequence of parts for removal from an EOL product that minimizes the resources for disassembly and maximizes the automation of the process and the quality of the parts or materials recovered. In Xanthopoulos and Iakovou (2009), lexicographic goal programming is employed to determine the most desirable components of an EOL product to be nondestructively disassembled. Ondemir and Gupta (2014b) develop a lexicographic mixed-integer goal programming model for an advanced remanufacturing-to-order and DTO system utilizing the life cycle data collected, stored, and delivered by the Internet of things.

Goal programming is employed in Gupta and Isaacs (1997) to investigate the effect of lightweighting on the dismantler and shredder's profitabilities associated with the EOL vehicle recycling industry of the United States. Isaacs and Gupta 1997) propose a goal programming-based methodology to explore changes to the current U.S. vehicle recycling infrastructure considering their effects on the dismantler and shredder's profitabilities. Boon et al. (2000); Boon et al. (2003) use goal programming to evaluate the materials streams and process profitabilities for several different aluminum-intensive vehicle (AIV) processing scenarios.

Gupta and Evans (2009) develop a nonpreemptive goal-programming model for operational planning of closed-loop supply chains considering multiple products and operations associated with the product, subassembly, part, and material levels.

Nasution et al. (2010) develop a goal-programming model to determine the most desirable disassembly process plan while satisfying various environmental, financial, and physical goals.

Harraz and Galal (2011) propose a goal programming-based methodology to solve a product recovery network design problem involving the determination of the locations for the different facilities and the amounts to be allocated to the different EOL options. Chaabane et al. (2011) develop a goal programming-based sustainable supply-chain design methodology by considering carbon emissions, suppliers and subcontractors selection, total logistics costs, technology acquisition, and the choice of transportation modes.

1.2.2 Fuzzy Goal Programming

Aspiration levels (goals) are considered concise and precise in goal programming. However, there are many occasions where a decision maker cannot specify the goal values precisely. Fuzzy goal programming takes this uncertainty into consideration by employing the concept of membership functions based on fuzzy set theory (Aouni et al. 2009).

Kongar et al. (2002) and Kongar and Gupta (2006) use fuzzy goal programming to determine the number of EOL products to be taken back, and the number of reused, recycled, stored, and disposed items.

Mehrbod et al. (2012) first develop a multiobjective mixed-integer nonlinear programming formulation for a closed-loop supply chain. Then, this model is solved using interactive fuzzy goal programming (IFGP), which has the ability to address the imprecise nature of decision makers' aspiration levels for goals. Ghorbani et al. (2014) develop a fuzzy goal-programming model for the design of a reverse logistics network.

Imtanavanich and Gupta (2005) employ weighted fuzzy goal programming to solve the multiperiod DTO problem that involves the disassembly of a variety of returned products to fulfill the demand for specified numbers of components and materials. Imtanavanich and Gupta (2006d) integrate genetic algorithms (GAs) with weighted fuzzy goal programming to solve a similar DTO problem.

1.2.3 Physical Programming

Physical programming uses a preference function to represent the decision maker's preference. In physical programming, the decision maker determines a suitable preference function and specifies ranges of different degrees of desirability (desirable, tolerable, undesirable, etc.) for each criterion. There are eight preference functions, classified into eight classes: four soft and four hard (Lambert and Gupta 2005; Ilgin and Gupta 2012a).

We can classify physical programming studies in ECMPRO into two subcategories, namely, reverse and closed-loop supply-chain network design and DTO systems.

1.2.3.1 Reverse and Closed-Loop Supply-Chain Network Design

In Pochampally et al. (2003), linear physical programming (LPP) is employed to identify potential facilities from a set of candidate recovery facilities operating in a region by considering several criteria (namely, quality of products at recovery facility, ratio of throughput to supply of used products, multiplication of throughput by disassembly time, and customer service rating of the recovery facility). Pochampally and Gupta (2004) develop an LPP-based reverse supply-chain design methodology involving three phases. Economic products to be reprocessed are selected from a set of candidate cores in Phase 1. Phase 2 involves the determination of potential recovery facilities using the criteria and classes defined in Pochampally et al. (2003). Phase 3 determines the right mix and quantities of products to be transported within the reverse supply chain. The strategic and tactical planning model developed by Nukala and Gupta (2006a) determines the following variables simultaneously: the most economic used product to reprocess, efficient production facilities, and the right mix and quantity of goods to be transported across the supply chain. Similar models are presented in Pochampally et al. (2008); Pochampally et al. (2009b) and Ilgin and Gupta (2012b).

Quality function deployment (QFD) and LPP are integrated in Pochampally et al. (2009a) to measure the *satisfaction level* of a reverse/closed-loop supply chain with respect to various performance measures such as reputation and innovation. Pochampally et al. (2009b) present a similar model.

An LPP-based methodology for collection center selection problem is presented in Pochampally and Gupta (2012), considering eight criteria [namely, sigma level (SL), per capita income of people in residential area (PR), utilization of incentives from local government (UG), distance from residential area (DR), distance from highways (DH), incentives from local government (IG), space cost (SC), labor cost (LC)].

1.2.3.2 Disassembly-to-Order Systems

Disassembly is a critical operation in product recovery process, since all product disposal options (e.g., recycling, remanufacturing) require the disassembly of EOL products at some level (Tang et al. 2002). Significant improvements can be achieved in the profitability of product recovery options by effectively planning the disassembly process. One of the important disassembly planning problems is the DTO problem, which involves the determination of the number of EOL products to be processed to fulfill a certain demand for products, parts and/or materials under a variety of objectives and constraints. An LPP model is developed in Kongar and Gupta (2002) to solve a DTO problem that involves the determination of the number of items to be disassembled for remanufacturing, recycling, storage, and disposal. The criteria considered include average customer satisfaction, average quality achievement, resale revenue, recycling revenue, total profit, number of recycled items, average environmental damage, average environmental benefit, and number of disposed items. Lambert and Gupta (2005) present a similar model. A DTO problem is modeled by Kongar and Gupta (2009) considering five goals (number of disposed items, total profit, number of recycled items, environmental damage, and customer satisfaction). In Imtanavanich and Gupta (2006c), LPP is used to solve a multiperiod DTO problem. GAs and LPP are integrated in Imtanavanich and Gupta (2006b) to solve a DTO problem. The fitness value of GAs is calculated using LPP. A multiperiod DTO problem with four objectives (i.e., maximization of profit, minimization of procurement cost, minimization of purchase cost, and minimization of disposal cost) is solved in Massoud and Gupta (2010a) by developing an LPP-based solution approach. Optimum disassembly, refurbishment, disposal, recycling, and storage plans are determined by the LPP model developed by Ondemir and Gupta (2011) for a demand-driven environment utilizing the life cycle data collected, stored, and delivered by sensors and radio-frequency identification (RFID) tags. Ondemir and Gupta (2014a) develop an LPP model to optimize a multicriteria advanced repair-to-order and DTO system involving sensor embedded products.

1.2.4 Data Envelopment Analysis

Data envelopment analysis (DEA) is used to evaluate the performance of a set of peer entities called *decision making units* (DMUs) that convert multiple inputs into multiple outputs (Cooper et al. 2011).

Kumar and Jain (2010) develop a DEA model of green supplier selection by considering carbon footprints of suppliers as a necessary dual-role factor. Mirhedayatian et al. (2014) evaluate the performance of green supply chains by developing a network DEA model involving dual-role factors, undesirable outputs, and fuzzy data.

The DEA-based methodology proposed by Saen (2009) determines the most efficient third-party reverse logistics provider (3PRLP) considering quantitative and qualitative data. Saen (2010) proposes a DEA-based 3PRLP selection methodology for the case of multiple dual factors while Saen (2011) and Azadi and Saen (2011) provide 3PL selection models involving both multiple dual factors and imprecise data. Zhou et al. (2012) develop a fuzzy confidence DEA model to select third-party recyclers.

1.2.5 Other Mathematical Models

Bouchery et al. (2012) reformulate the classical economic order quantity model as a multiobjective problem and call it a *sustainable order quantity model*. They also considered a multiechelon extension of this model. The set of efficient solutions (Pareto optimal solutions) is analytically characterized for both models. In addition, an interactive procedure helping decision makers in the quick identification of the best option among these solutions is proposed.

Humphreys et al. (2006) use dynamic fuzzy membership functions to select green supplies. Feyzioglu and Büyüközkan (2010) employ 2-additive Choquet integrals to consider criteria dependencies in green supplier evaluation.

Wang et al. (2011) develop a multiobjective mixed-integer programming formulation for a green supply-chain network design problem by considering the trade-off between the total cost and the environment influence. Pishvaee and Razmi (2012) design an environmental supply chain under uncertainty using multiobjective fuzzy mathematical programming. Samanlioglu (2013) proposes a multiobjective mixed-integer model for the location-routing decisions of industrial hazardous material management. Ramezani et al. (2013) present a stochastic multiobjective model for the design of a forward/reverse supply-chain network with the goals of maximization of profit, maximization of responsiveness, and minimization of defective parts from suppliers. Özkır and Başlıgil (2013) propose a fuzzy multiobjective optimization model for the design of a closed-loop supply-chain network. The mixed-integer programming model proposed by Ozceylan and Paksoy (2013b) determines the optimum transportation amounts together with the location of plants and retailers by considering multiple periods and multiple parts. Ozceylan and

Paksoy (2013a) develop a fuzzy multiobjective linear-programming model for the design of a closed-loop supply chain by considering the uncertainty associated with capacity, demand, and reverse rates. Mirakhorli (2014) proposes an interactive fuzzy multiobjective linear-programming model to solve a fuzzy biobjective reverse logistics network design problem. Nurjanni et al. (2014) integrate three scalarization approaches, namely, the weighted sum method, the weighted Tchebycheff method, and the augmented weighted Tchebycheff method, to solve the mathematical model associated with a closed-loop supply-chain network.

1.3 Qualitative Techniques

While the quantitative techniques consider continuous decision spaces, qualitative techniques concentrate on problems with discrete decision spaces (Roostaee et al. 2012). They consider a limited number of predetermined alternatives and discrete preference ratings (Tzeng and Huang 2011). In this section, we divide the ECMPRO applications of qualitative techniques into thirteen parts: analytical hierarchy process (AHP), fuzzy AHP, analytical network process (ANP), DEMATEL, TOPSIS, ELECTRE, PROMETHEE, multiattribute utility theory (MAUT), VIKOR, MACBETH, case-based reasoning (CBR), gray relational analysis (GRA), and other techniques.

1.3.1 Analytical Hierarchy Process

AHP is an MCDM tool formalized by Saaty (1980). It uses simple mathematics to support decision makers in explicitly weighing tangible and intangible criteria against each other for the purpose of resolving conflict or setting priorities.

Azzone and Noci (1996) use AHP to evaluate the environmental performance of alternative product designs. In Choi et al. (2008), the relative importance of five design-for-environment strategies are compared using AHP. Wang et al. (2012) develop an AHP-based green product design selection methodology that does not require the designers to conduct detailed analysis (e.g., life cycle assessment) for every new product option. Kim et al. (2009) employ AHP to evaluate the recycling potential of materials based on environmental and economic factors.

Noci (1997) proposes a green vendor rating system using AHP. Handfield et al. (2002) develop an AHP-based methodology to assess the relative performance of several suppliers considering environmental issues. Dai and Blackhurst (2012) develop a four-phase methodology for sustainable supplier assessment by integrating QFD and AHP. First, customer requirements are linked with the company's sustainability strategy. Then, the sustainable purchasing competitive priority is determined. Next, sustainable supplier assessment criteria are developed. Finally, AHP is employed to assess the suppliers. Shaik and Abdul-Kader (2012) first develop a reverse logistics performance measurement system that is based on balanced scorecard and performance prism. Then, AHP is integrated with this system to calculate the overall comprehensive performance index (OCPI).

Barker and Zabinsky (2011) use sensitivity analysis with AHP to provide insights into the preference ordering among eight alternative reverse logistics network configurations. In Jiang et al. (2012), AHP is used for remanufacturing portfolio selection. Ziout et al. (2013) develop an AHP-based methodology to evaluate the sustainability level of manufacturing systems. The AHP-based methodology proposed by Sarmiento and Thomas (2010) identifies improvement areas in the implementation of green initiatives. Subramoniam et al. (2013) use AHP to validate the Reman decision making framework (RDMF) developed in Subramoniam et al. (2010).

1.3.2 Fuzzy Analytical Hierarchy Process

There are two characteristics of AHP often criticized in the literature: the use of an unbalanced scale of judgments and the inability to adequately handle the inherent uncertainty and imprecision in the pairwise comparison process (Ertugrul and Karakasoglu 2009). A fuzzy analytical process that integrates AHP with the concepts of fuzzy set theory is often used by researchers to overcome these limitations of AHP.

AHP and fuzzy multiattribute decision making are integrated in the environmentally conscious design methodology proposed by Kuo et al. (2006). Li et al. (2008) integrate AHP and fuzzy logic to determine an optimal modular formulation in modular product design with environmental considerations.

Yu et al. (2000) use fuzzy AHP to determine the most appropriate recycling option for EOL products considering three criteria: environmental impact, recycling associated cost, and recoverable material content.

In Lu et al. (2007), Lee et al. (2009), Grisi et al. (2010), Çiftçi and Büyüközkan (2011), and Amin and Zhang (2012), fuzzy AHP is used to integrate environmental factors into the supplier evaluation process. Lee et al. (2012) propose a fuzzy AHP-based approach to determine the most important criteria for green supplier selection in the Taiwanese hand tool industry. In Chiou et al. (2008), fuzzy AHP is employed to compare the green supply-chain management (GSCM) practices of American, Japanese, and Taiwanese electronics manufacturers operating in China. Chiou et al. (2012) employ fuzzy AHP to select the most important criteria in reverse logistics implementation. Efendigil et al. (2008) present an approach integrating fuzzy AHP and artificial neural networks for the third-party reverse logistics provider selection problem.

Gupta and Nukala (2005) use fuzzy AHP to identify potential facilities in a set of candidate recovery facilities operating in the region. Shaverdi et al. (2013) employ fuzzy AHP to determine the effective factors associated with the sustainable supply-chain management in the publishing industry.

1.3.3 Analytical Network Process

ANP was developed by Saaty (1996) as a generalization of AHP. It releases the restrictions of the hierarchical AHP structure by modeling the decision problem as an influence network of clusters and nodes contained within the clusters.

ANP is used in Cheng and Lee (2010) to investigate the relative importance of service requirements as well as selecting an appropriate third-party reverse logistics provider. Meade and Sarkis (2002) employ ANP for the evaluation and selection of third-party reverse logistics providers. Hsu and Hu (2007); Hsu and Hu (2009) integrate hazardous substance management to supplier selection using ANP. Büyüközkan and Çiftçi (2011, 2012) use fuzzy ANP to evaluate GSCM practices of an automotive company and propose a fuzzy ANP-based methodology for sustainable supplier selection.

Ravi et al. (2005) use ANP together with balanced score card to determine the most suitable EOL option for EOL computers. Chen et al. (2012) solve the GSCM strategy selection problem of a Taiwanese electronics company using ANP.

Sarkis (1998) employs ANP to evaluate environmentally conscious business practices. In Vinodh et al. (2012), the environmentally conscious business practice model proposed by Sarkis (1998) is adopted for the evaluation of sustainable business practices in an Indian relay manufacturing organization. Chen et al. (2009) use ANP to evaluate several GSCM strategies (i.e., green design, green purchasing, green marketing, green manufacturing). Bhattacharya et al. (2014) develop an intraorganizational collaborative decision making (CDM) approach for performance measurement of a green supply chain (GSC) by integrating fuzzy ANP and balanced score card. Tuzkaya and Gulsun (2008) integrate ANP with fuzzy TOPSIS to evaluate centralized return centers in a reverse logistics network.

Gungor (2006) develops an ANP-based methodology to evaluate connection types in design for disassembly.

1.3.4 DEMATEL

The decision making and evaluation laboratory (DEMATEL) method is used to identify causal relationships among the elements of a system. The main output of this technique is a causal diagram that uses digraphs instead of directionless graphs to describe the contextual relationships and the strengths of influence among the elements (Wu 2008).

Lin (2013) uses fuzzy DEMATEL to analyze the interrelationships among three issues (GSCM practices, organizational performance, and external driving factors) associated with GSCM implementation.

1.3.5 TOPSIS

The technique for order preference by similarity to ideal solution (TOPSIS) determines the best alternative based on the concept of the compromise solution that is the shortest distance from the ideal solution and the greatest distance from the negative-ideal solution in a Euclidean sense (Tzeng and Huang 2011).

Gupta and Pochampally (2004) propose a fuzzy TOPSIS-based approach for the evaluation of recycling programs with respect to drivers of public participation. Remery et al. (2012) propose a TOPSIS-based EOL option selection methodology called *ELSEM*, while Wadhwa et al. (2009) use fuzzy TOPSIS for the option selection problem in reverse logistics. Gao et al. (2010) construct a fuzzy TOPSIS model to evaluate a set of feasible green design alternatives. A fuzzy TOPSIS approach is proposed in Yeh and Xu (2013) for the evaluation of alternative recycling activities of a recycling company considering various sustainability criteria with environmental, economic, and social dimensions. Vinodh et al. (2013) use fuzzy TOPSIS to determine the best sustainable concept among five sustainable concepts (i.e., design for environment, life cycle assessments, environmentally conscious QFD, theory of inventive problem solving, and life cycle impact assessment). Mahapatara et al. (2013) develop a fuzzy TOPSIS methodology to evaluate different reverse manufacturing alternatives (remanufacturing, reselling, repairing, cannibalization, and refurbishing). Diabat et al. (2013) develop a fuzzy TOPSIS-based methodology to assess the importance of GSCM practices and performances in an automotive company.

Kannan et al. (2009) integrate interpretive structural modeling and fuzzy TOPSIS to select the best third-party reverse logistics provider. Awasthi et al. (2010), Govindan et al. (2012), and Shen et al. (2013) use fuzzy TOPSIS to generate an overall performance score to measure the environmental performance of suppliers.

1.3.6 ELECTRE

ELimination Et Choix Traduisant la REalité (ELECTRE) (in French), which means elimination and choice expressing reality, performs pairwise comparisons among alternatives for each one of the attributes separately to establish outranking relationships between the alternatives (Bari Leung 2007). These outranking relations are built in such a way that it is possible to compare alternatives. The information required by ELECTRE consists of information among the criteria and information within each criterion (Teixeira 2007).

Bufardi et al. (2004) employ ELECTRE III for the selection of the best EOL alternative.

1.3.7 PROMETHEE

The preference ranking organization method for enrichment evaluation (PROMETHEE) is a prescriptive method that enables a decision maker to rank the alternatives according to his/her preferences. It requires a preference function associated with each criterion, as well as weights indicating their relative importance. While PROMETHEE I gives a partial ranking of alternatives, PROMETHEE II gives a complete ranking Brans and Mareschal (2005); Mareschal and Smet (2009).

Avikal et al. (2013b) develop a PROMETHEE-based methodology for assigning the disassembly tasks to workstations of a disassembly line. Ghazilla et al. (2013) use PROMETHEE to evaluate alternative fasteners in design for disassembly.

1.3.8 Multiattribute Utility Theory (MAUT)

In MAUT, the decision maker represents a complex problem as a simple hierarchy and subjectively evaluates a large number of quantitative and qualitative factors considering risk and uncertainty. MAUT can be used in both deterministic and stochastic decision environments (Min 1994).

Erol et al. (2011) integrate fuzzy entropy and fuzzy multiattribute utility (FMAUT) to measure the sustainability performance of a supply chain. First, the fuzzy entropy method is used to determine the importance levels for the indicators. Then, FMAUT is utilized to calculate the aggregated performance indices with respect to each aspect of sustainability. Shaik and Abdul-Kader (2011) present the use of MAUT to develop an integrated and comprehensive framework for green supplier selection by considering traditional aspects as well as environmental and social factors.

1.3.9 VIKOR

VlseKriterijumskaOptimizacija I KompromisnoResenje (VIKOR) (in Serbian), which means multicriteria optimization and compromise solution method, determines the compromise ranking list, the compromise solution, and the weight stability intervals for preference stability of the compromise solution obtained with the initial (given) weights. It is especially useful when there are conflicting criteria in the decision problem (Opricovic and Tzeng 2004).

Rao (2009) proposes a VIKOR-based methodology for the selection of environmentally conscious manufacturing programs.

The green supplier selection and evaluation methodology developed by Datta et al. (2012) integrates VIKOR with an interval-valued fuzzy set.

Samantra et al. (2013) use the methodology proposed in Datta et al. (2012) to determine the best product recovery option.

Sasikumar and Haq (2011) propose a two-step methodology for the design of a closed-loop supply chain. First, VIKOR is used to select the best 3PRLP. Then, a mixed-integer linear-programming model is developed to make decisions on raw material procurement, production, and distribution.

1.3.10 MACBETH

Measuring attractiveness by a categorical based evaluation technique (MACBETH) is a technique similar to AHP. The only difference is that MACBETH uses an interval scale while AHP adopts a ratio scale (Ishizaka and Nemery 2013).

Dhouib (2014) proposes a fuzzy MACBETH methodology to evaluate options in reverse logistics for waste automobile tires.

1.3.11 Case-Based Reasoning

CBR is based on a memory-centered cognitive model. In this method, a reasoner remembers a previous situation similar to the current one and uses that to solve the new problem (Xu 1994; Kolodner 1992).

Zeid et al. (1997) present a CBR-based methodology to determine the disassembly plan of a single product. Extending Zeid et al. (1997), Gupta and Veerakamolmal (2000) and Veerakamolmal and Gupta (2002) develop CBR approaches to automatically generate disassembly plans for multiple products.

Humphreys et al. (2003) consider environmental factors in the supplier selection process by developing a knowledge-based system (KBS) that integrates CBR and decision support components.

1.3.12 Gray Relational Analysis

In GRA, simple mathematical relations are used to deal with uncertain, poor, and incomplete information. GRA solves multiattribute decision making problems by combining the entire range of performance attribute values being considered for every alternative into one single value (Kuo et al. 2008).

Chan (2008) employs GRA to rank the product EOL options under uncertainty with respect to several criteria at the material level. Li and Zhao (2009) integrate the threshold method and GRA for the selection of green suppliers. In Chen et al. (2010), fuzzy logic and GRA are integrated to determine suitable suppliers by considering various environment-related criteria.

1.3.13 Other Techniques

Rao and Padmanabhan (2010) use digraph and matrix methods for the selection of the best product EOL scenario. Bereketli et al. (2011) evaluate alternative waste treatment strategies for electrical and electronic equipment using the fuzzy linear-programming technique for multidimensional analysis of preference (LINMAP). Yang and Wu (2007) employ the gray entropy method for the green supplier selection problem, while Yu-zhong and Li-yun (2008) solve the same problem using the extension method based on entropy weight. Iakovou et al. (2009) develop a multicriteria analysis technique called the *multicriteria matrix* to rank components according to their potential value at the end of their useful life. In Lee et al. (2001), a multiobjective methodology has been developed to determine an appropriate EOL option for a product. Sangwan (2013) develop a multicriteria performance analysis tool to evaluate the performance of manufacturing systems based on environmental criteria. Mangla et al. (2014) use interpretive structural modeling to analyze the interaction among the GSC variables.

1.4 Mixed Techniques

The complex and interdisciplinary nature of ECMPRO-related problems often requires the integration of two or more multicriteria optimization approaches. This section presents an overview of these integrated approaches.

1.4.1 Analytical Hierarchy Process and Data Envelopment Analysis

Wen and Chi (2010) integrate AHP/ANP with DEA to develop a green supplier selection procedure. First, DEA distinguishes the efficient supplier candidates from the entire group. Then, AHP/ANP is used for further analysis without making efforts to deal with inefficient suppliers.

1.4.2 PROMETHEE and Goal Programming

Walther et al. (2006) present a two-step methodology for the evaluation of alternative scrap treatment systems. First, linear programming or weighted goal programming is used to determine short-term decisions. Then, the results obtained in the first step are used as *a priori* information for the multicriteria decision making tool PROMETHEE at strategic level.

1.4.3 PROMETHEE and Analytical Hierarchy Process

Avikal et al. (2013a) solve the disassembly line balancing problem by developing an AHP/TOPSIS-based methodology. In the proposed heuristic, the important criteria, which play a significant role in the product disassembly process, are selected. Then, AHP is applied to calculate the weight of each criterion. Finally, PROMETHEE uses these weights to determine the ranking of the tasks for the assignment to the disassembly stations. Avikal et al. (2014) modify Avikal et al.'s (2013a) methodology by using fuzzy AHP instead of AHP.

1.4.4 PROMETHEE and Analytical Network Process

Tuzkaya et al. (2009) evaluate the environmental performance of suppliers by developing a methodology that integrates fuzzy ANP and fuzzy PROMETHEE.

1.4.5 Analytical Hierarchy Process and Case-Based Reasoning

Kuo (2010) integrates AHP and CBR to simplify the calculation of the recyclability index, which is used to evaluate the recyclability of an EOL product. Ghazalli and Murata (2011) integrate AHP and CBR to evaluate EOL options for parts and components.

1.4.6 Analytical Network Process and Goal Programming

Nukala and Gupta (2006b) employ ANP/goal programming integration for the supplier selection problem of a closed-loop supply chain. First, the supply-chain strategy is determined qualitatively by evaluating the suppliers with respect to several criteria. Then, taking ANP ratings as input, preemptive goal programming is used to determine the optimal quantities to be ordered from the suppliers.

In Ravi et al. (2008), an integrated ANP/goal programming methodology is used to select reverse logistics projects. Following the determination of the level of interdependence among the criteria and candidate reverse logistics projects using ANP, zero-one goal programming determines the allocation of resources among reverse logistics projects by considering resource limitations and several other selection constraints.

1.4.7 Analytical Network Process and Data Envelopment Analysis

Sarkis (1999) integrates ANP and DEA to evaluate environmentally conscious manufacturing programs. Kuo and Lin (2012) develop a methodology by coupling ANP and DEA for green supplier selection.

1.4.8 Analytical Hierarchy Process and Genetic Algorithms

Dehghanian and Mansour (2009) integrate AHP and GAs for the recovery network design of scrap tires. In the proposed methodology, first, AHP is used to calculate social impacts. Then, the Pareto optimal solutions are determined by using a multiobjective genetic algorithm (MOGA).

Vadde et al. (2011) analyze the pricing decisions of product recovery facilities by integrating multiobjective mathematical programming, GAs, and AHP. The weights used in the objective function of the GA designed to solve the multiobjective mathematical programming model are determined using AHP. Ge (2009) integrates GAs and AHP for the evaluation of green suppliers.

1.4.9 Analytical Hierarchy Process and Neural Networks

Thongchattu and Siripokapirom (2010) model the green supplier selection problem using AHP. Neural networks are used to determine criteria weights.

1.4.10 Analytical Hierarchy Process and Analytical Network Process

Govindan et al. (2013) develop a two-phase model for the selection of third-party reverse logistics providers. In this model, AHP is employed to identify the most prioritized factors while ANP is used to select the reverse logistic providers.

1.4.11 Analytical Hierarchy Process and TOPSIS

Wittstruck and Teuteberg (2012) integrate fuzzy AHP and TOPSIS for recycling partner selection. Senthil et al. (2012) determine the best reverse logistics operating channel by combining AHP and fuzzy TOPSIS. Wang and Chan (2013) integrate fuzzy TOPSIS and AHP for the evaluation of new green initiatives. In Ravi (2012) and Senthil et al. (2014), AHP/TOPSIS integration is employed for the selection of third-party reverse logistics providers.

1.4.12 Analytical Network Process and Gray Relational Analysis

In Dou et al. (2014), ANP and GRA are integrated to determine effective green supplier development programs.

1.4.13 Analytical Hierarchy Process and Simulation

De Felice and Petrillo (2012) integrate AHP and simulation to simultaneously improve the performance of inventory management and reverse logistics management.

1.4.14 Analytical Hierarchy Process and Structural Equation Modeling

The approach proposed by Punniyamoorty et al. (2012) combines AHP and structural equation modeling (SEM) for the selection of suppliers considering economic as well as environmental factors.

1.4.15 Approaches Involving More than Two Techniques

Pochampally and Gupta (2008) develop a three-phase methodology for the effective design of a reverse supply-chain network. The most economical product to reprocess from a set of different used products is selected in Phase 1 using a fuzzy benefit function. AHP and fuzzy set theory are employed in Phase 2 to identify potential facilities in a set of candidate recovery facilities. Phase 3 solves a single-period and single-product discrete location model to minimize overall cost across the reverse supply-chain network.

Nukala and Gupta (2007) integrate Taguchi loss functions, AHP, and fuzzy programming to evaluate the suppliers and determine the order quantities in a closed-loop supply-chain network.

First, Taguchi loss functions quantify the suppliers' attributes to quality loss. Then, AHP is used to transform these quality losses into a variable that is used in the formulation of the fuzzy programming objective function. Finally, fuzzy programming determines the order quantities.

Büyüközkan and Berkol (2011) integrate ANP, goal programming, and QFD to design a sustainable supply chain. ANP is employed to determine the importance levels in the house of quality by considering the interrelationships among the design requirements and customer requirements, while zero-one goal programming is used to select the most suitable design requirements based on ANP results.

Paksoy et al. (2012) first propose a fuzzy programming model with multiple objectives for the design of a closed-loop supply-chain network. Then, various multicriteria techniques (i.e., AHP, fuzzy AHP, and fuzzy TOPSIS) are applied to weight the objectives and the corresponding results are discussed.

Zareinejad and Javanmard (2013) develop an integrated methodology for third-party reverse logistics provider selection. First, relationships among the attributes are analyzed using ANP. Then, intuitionistic fuzzy set (IFS) and GRA are integrated to determine the most suitable third-party reverse logistics provider under uncertain conditions.

Hsu et al. (2011) develop a balanced score card to measure sustainable performance in the semiconductor industry. The fuzzy Delphi method and ANP are used to identify related measures and perspectives of sustainable balanced score card activities.

Hsu et al. (2012) combine DEMATEL, ANP, and VIKOR to solve the recycled material vendor selection problem. First, DEMATEL and ANP are integrated to determine the degrees of influence among the criteria. Then, VIKOR is employed to rank the alternative vendors.

Kannan et al. (2013) propose an integrated approach for supplier selection and order allocation in a GSC. First, the relative weights of supplier selection criteria are calculated using fuzzy AHP; then, fuzzy TOPSIS is employed to rank suppliers based on the selected criteria. Finally, the Multi-Objective Linear Programming (MOLP) model determines the optimal order quantity from each supplier using the weights of the criteria and ranks of suppliers.

1.5 Heuristics and Metaheuristics

A heuristic can be defined as a technique that seeks or finds good solutions to a difficult model. A metaheuristic extends the heuristic concept by exploiting ideas and concepts from another discipline to help solve the artificial system being modeled. GAs, simulated annealing, and tabu search are the most commonly used metaheuristics (Jones et al. 2002). In this section, we provide an overview of multiobjective heuristic and metaheuristic approaches developed to solve ECMPRO-related problems.

Gupta and Taleb (1996) and Taleb and Gupta (1996) presented a heuristic methodology for disassembling multiple product structures with parts/materials commonality. There are two companion algorithms in this methodology: the *core algorithm* and the *allocation algorithm*. The total disassembly requirements of the root items over the planning horizon are determined by the core algorithm and the schedule for disassembling the roots and the subassemblies are provided by the allocation algorithm. Langella (2007) extends this methodology by considering holding costs and external procurement of items.

GAs are numerical optimization algorithms inspired by both natural selection and natural genetics. They are generally used to search large, nonlinear search spaces where expert knowledge is lacking or difficult to encode and where traditional optimization methods fall short (Goldberg 1989). GAs are by far the most frequently used metaheuristic to solve ECMPRO-related problems. Jun et al. (2007) develop a multiobjective evolutionary algorithm to determine the best EOL option. In Hula et al. (2003), multiobjective GAs are used to determine the most appropriate EOL option. The GSC partner selection problem is solved in Yeh and Chuang (2011) by developing two multiobjective GAs. The multiobjective GA developed by Sakundarini et al. (2013) considers technical, economic, and recyclability requirements for the selection of materials with high recyclability. Rickli and Camelio (2013) develop a multiobjective genetic algorithm to optimize partial disassembly sequences based on disassembly operation costs, recovery reprocessing costs, revenues, and environmental impacts. Chern et al. (2013) develop a heuristic called the *genetic algorithms-based master planning algorithm* (GAMPA) that solves the master planning problem of a supply-chain network involving multiple final products, substitutions, and a recycling process. Liu and Huang (2014) use multiobjective genetic algorithms to solve two scheduling problems involving economic and environment-related criteria. Wang et al. (2014) present an application of multiobjective GAs in closed-loop supply-chain network design.

Besides GAs, researchers have applied several other metaheuristics to ECMPRO-related problems. Guo et al. (2012) propose a multiobjective scatter search algorithm to solve the selective disassembly problem that involves the determination of optimal disassembly sequences for single or multiple target components. Jamshidi et al. (2012) develop a mathematical model for the design of a supply chain by simultaneously considering cost and environmental effect. A memetic algorithm integrated with the Taguchi method is utilized to solve the model.

The disassembly line balancing problem (DLBP) is an important and actively researched problem in ECMPRO (see McGovern and Gupta [2011] for more information on DLBP). It is a multiobjective problem, as described by Gungor and Gupta (2002), and has been mathematically proven to be NP-complete by McGovern and Gupta (2007), which makes the desire to achieve the best balance computationally expensive when considering large-sized problems. Thus, the need to obtain near-optimal solutions efficiently have led various authors to use a variety of heuristic and metaheuristic techniques such as

- Genetic algorithms (GAs) (McGovern and Gupta 2007; Aydemir-Karadag and Turkbey 2013)
- Ant colony optimization (ACO) (Agrawal and Tiwari 2008; Kalayci and Gupta 2013a)
- Simulated annealing (SA) (Kalayci and Gupta 2013c)
- Tabu search (TS) (Kalayci and Gupta 2014)
- Artificial bee colony (ABC) (Kalayci and Gupta 2013b)
- Particle swarm optimization (PSO) (Kalayci and Gupta 2013d)
- River formation dynamics (RFD) (Kalayci and Gupta 2013e)

Delorme et al. (2014) present an overview of multiobjective approaches developed for the design of assembly and disassembly lines.

1.6 Simulation

Simulation is generally employed to analyze complex processes or systems. It involves the development and analysis of models that have the ability to imitate the behavior of the system being analyzed (Pegden and Shannon 1995).

Shokohyar and Mansour (2013) deal with the electronic waste management problem of Iran by developing a simulation–optimization model that determines the locations for collection centers and recycling plants. A fuzzy controlled agent-based simulation framework proposed by Zhang et al. (2003) evaluates the environmental performance of the suppliers.

1.7 Conclusions

In this chapter, we presented an overview of the state of the art on the use of MCDM techniques in ECMPRO (Ilgin et al. 2015). The reviewed studies were classified into five categories (i.e., quantitative techniques, qualitative techniques, mixed techniques, heuristics and metaheuristics, and simulation). The following general conclusions can be drawn from our literature review:

- Qualitative techniques are by far the most frequently used MCDM techniques (see Figure 1.2). Among qualitative techniques, AHP, ANP, and TOPSIS are the most popular ones. On the other hand, the use of some qualitative techniques (i.e., MACBETH, DEMATEL, ELECTRE, PROMETHEE) is very rare. In addition, there has been no application of some qualitative MCDM techniques (e.g., MOORA, COPRAS) in ECMPRO yet.
- There is a significant increase in the number of studies applying MCDM techniques to ECMPRO problems in recent years. This can be attributed to the increasing popularity of environmental issues among researchers.

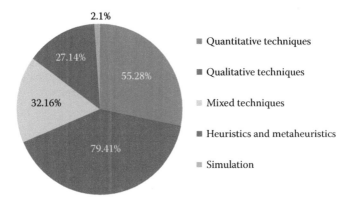

FIGURE 1.2
Number and percentage of publications in each category.

- Green supplier selection and evaluation, disassembly planning, reverse logistics, and closed-loop supply-chain design are the major ECMPRO issues analyzed and solved by using MCDM techniques.
- Although simulation is very good at modeling complex systems, its integration with MCDM techniques for the solution of ECMPRO-related problems is limited to a few studies. Hence, there are opportunities to develop multiobjective solution methodologies integrating simulation with qualitative and/or quantitative MCDM techniques to solve complex ECMPRO issues such as disassembly planning, reverse logistics, and closed-loop supply-chain design.

References

Agrawal S, Tiwari MK. A collaborative ant colony algorithm to stochastic mixed-model U-shaped disassembly line balancing and sequencing problem. *International Journal of Production Research* 2008;46:1405–29.

Amin SH, Zhang G. An integrated model for closed-loop supply chain configuration and supplier selection: Multi-objective approach. *Expert Systems with Applications* 2012;39:6782–91.

Aouni B, Martel J-M, Hassaine A. Fuzzy goal programming model: An overview of the current state-of-the art. *Journal of Multi-Criteria Decision Analysis* 2009;16:149–61.

Avikal S, Mishra PK, Jain R. An AHP and PROMETHEE methods-based environment friendly heuristic for disassembly line balancing problems. *Interdisciplinary Environmental Review* 2013a;14:69–85.

Avikal S, Mishra PK, Jain R. A fuzzy AHP and PROMETHEE method-based heuristic for disassembly line balancing problems. *International Journal of Production Research* 2014;52:1306–17.

Avikal S, Mishra PK, Jain R, Yadav HC. A PROMETHEE method based heuristic for disassembly line balancing problem. *Industrial Engineering & Management Systems* 2013b;12:254–63.

Awasthi A, Chauhan SS, Goyal SK. A fuzzy multicriteria approach for evaluating environmental performance of suppliers. *International Journal of Production Economics* 2010;126:370–8.

Aydemir-Karadag A, Turkbey O. Multi-objective optimization of stochastic disassembly line balancing with station paralleling. *Computers and Industrial Engineering* 2013;65:413–25.

Azadi M, Saen RF. A new chance-constrained data envelopment analysis for selecting third-party reverse logistics providers in the existence of dual-role factors. *Expert Systems with Applications* 2011;38:12231–6.

Azzone G, Noci G. Measuring the environmental performance of new products: An integrated approach. *International Journal of Production Research* 1996;34:3055–78.

Bari F, Leung V. Application of ELECTRE to network selection in a hetereogenous wireless network environment, in: IEEE *Wireless Communications and Networking Conference*, Kowloon, Hong Kong, 2007, pp. 3810–15.

Barker TJ, Zabinsky ZB. A multicriteria decision making model for reverse logistics using analytical hierarchy process. *Omega* 2011;39:558–73.

Bereketli I, Erol Genevois M, Esra Albayrak Y, Ozyol M. WEEE treatment strategies-evaluation using fuzzy LINMAP method. *Expert Systems with Applications* 2011;38:71–9.

Bhattacharya A, Mohapatra P, Kumar V, Dey PK, Brady M, Tiwari MK, Nudurupati SS. Green supply chain performance measurement using fuzzy ANP-based balanced scorecard: A collaborative decision-making approach. *Production Planning & Control* 2014;25:698–714.

Boon JE, Isaacs JA, Gupta SM. Economic impact of aluminum-intensive vehicles on the U.S. automotive recycling infrastructure. *Journal of Industrial Ecology* 2000;4:117–34.

Boon JE, Isaacs JA, Gupta SM. End-of-life infrastructure economics for "clean vehicles" in the United States. *Journal of Industrial Ecology* 2003;7:25–45.

Bouchery Y, Ghaffari A, Jemai Z, Dallery Y. Including sustainability criteria into inventory models. *European Journal of Operational Research* 2012;222:229–40.

Brans J-P, Mareschal B, Promethee methods, in: *Multiple Criteria Decision Analysis: State of the Art Surveys*, Springer New York, 2005, pp. 163–86.

Bufardi A, Gheorghe R, Kiritsis D, Xirouchakis P. Multicriteria decision-aid approach for product end-of-life alternative selection. *International Journal of Production Research* 2004;42:3139–57.

Buyukozkan G, Berkol C. Designing a sustainable supply chain using an integrated analytic network process and goal programming approach in quality function deployment. *Expert Systems with Applications* 2011;38:13731–48.

Büyüközkan G, Çiftçi G. A novel fuzzy multi-criteria decision framework for sustainable supplier selection with incomplete information. *Computers in Industry* 2011;62:164–74.

Büyüközkan G, Çiftçi G. Evaluation of the green supply chain management practices: A fuzzy ANP approach. *Production Planning & Control* 2012;23:405–18.

Chaabane A, Ramudhin A, Paquet M. Designing supply chains with sustainability considerations. *Production Planning & Control* 2011;22:727–41.

Chan JWK. Product end-of-life options selection: Grey relational analysis approach. *International Journal of Production Research* 2008;46:2889–912.

Chen C-C, Shih H-S, Shyur H-J, Wu K-S. A business strategy selection of green supply chain management via an analytic network process. *Computers & Mathematics with Applications* 2012;64:2544–57.

Chen CC, Shih H-S, Wu K-S, Shyur H-J. Using ANP for the selection of green supply chain management strategies, in: *Proceedings of the 10th International Symposium on the Analytic Hierarchy/Network Process*, Pittsburgh, PA, 2009.

Chen CC, Tseng ML, Lin YH, Lin ZS. Implementation of green supply chain management in uncertainty, in: *IEEE International Conference on Industrial Engineering and Engineering Management*, 2010, pp. 260–4.

Cheng Y-H, Lee F. Outsourcing reverse logistics of high-tech manufacturing firms by using a systematic decision-making approach: TFT-LCD sector in Taiwan. *Industrial Marketing Management* 2010;39:1111–19.

Chern C-C, Lei S-T, Huang K-L. Solving a multi-objective master planning problem with substitution and a recycling process for a capacitated multi-commodity supply chain network. *Journal of Intelligent Manufacturing* 2013;25:1–25.

Chiou CY, Chen HC, Yu CT, Yeh CY. Consideration factors of reverse logistics implementation: A case study of Taiwan's electronics industry. *Procedia: Social and Behavioral Sciences* 2012;40:375–81.

Chiou CY, Hsu CW, Hwang WY. Comparative investigation on green supplier selection of the American, Japanese and Taiwanese electronics industry in China, in: *IEEE International Conference on Industrial Engineering and Engineering Management*, Singapore, 2008, pp. 1909–14.

Choi JK, Nies LF, Ramani K. A framework for the integration of environmental and business aspects toward sustainable product development. *Journal of Engineering Design* 2008;19:431–46.

Ciftci G, Buyukozkan G. A fuzzy MCDM approach to evaluate green suppliers. *International Journal of Computational Intelligence Systems* 2011;4:894–909.

Cooper WW, Seiford LM, Zhu J, Cooper WW, Seiford LM. Data envelopment analysis: History, models, and interpretations, in: *Handbook on Data Envelopment Analysis*, Springer US, 2011, pp. 1–39.

Dai J, Blackhurst J. A four-phase AHP-QFD approach for supplier assessment: A sustainability perspective. *International Journal of Production Research* 2012;50:5474–90.

Datta S, Samantra C, Mahapatra SS, Banerjee S, Bandyopadhyay A. Green supplier evaluation and selection using VIKOR method embedded in fuzzy expert system with interval-valued fuzzy numbers. *International Journal of Procurement Management* 2012;5:647–78.

De Felice F, Petrillo A. Hierarchical model to optimize performance in logistics policies: Multi attribute analysis. *Procedia: Social and Behavioral Sciences* 2012;58:1555–64.

Dehghanian F, Mansour S. Designing sustainable recovery network of end-of-life products using genetic algorithm. *Resources, Conservation and Recycling* 2009;53:559–70.

Delorme X, Battaia O, Dolgui A. Multi-objective approaches for design of assembly lines, in: L. Benyoucef, J.-C. Hennet, M.K. Tiwari (Eds.) *Applications of Multi-Criteria and Game Theory Approaches*, Springer, New York, 2014, pp. 31–56.

Dhouib D. An extension of MACBETH method for a fuzzy environment to analyze alternatives in reverse logistics for automobile tire wastes. *Omega* 2014;42:25–32.

Diabat A, Khodaverdi R, Olfat L. An exploration of green supply chain practices and performances in an automotive industry. *The International Journal of Advanced Manufacturing Technology* 2013;68:949–61.

Dou Y, Zhu Q, Sarkis J. Evaluating green supplier development programs with a grey-analytical network process-based methodology. *European Journal of Operational Research* 2014;233:420–31.

Efendigil T, Onut S, Kongar E. A holistic approach for selecting a third-party reverse logistics provider in the presence of vagueness. *Computers & Industrial Engineering* 2008;54:269–87.

Erol I, Sencer S, Sari R. A new fuzzy multi-criteria framework for measuring sustainability performance of a supply chain. *Ecological Economics* 2011;70:1088–100.

Ertugrul I, Karakasoglu N. Performance evaluation of Turkish cement firms with fuzzy analytic hierarchy process and TOPSIS methods. *Expert Systems with Applications* 2009;36:702–15.

Feyzioglu O, Büyüközkan G. Evaluation of green suppliers considering decision criteria dependencies, in: M. Ehrgott, B. Naujoks, T.J. Stewart, and J. Wallenius (Eds.) *Multiple Criteria Decision Making for Sustainable Energy and Transportation Systems*, Springer, 2010, pp. 145–54.

Gao Y, Liu Z, Hu D, Zhang L, Gu G. Selection of green product design scheme based on multi-attribute decision-making method. *International Journal of Sustainable Engineering* 2010;3:277–91.

Ge Y. Research on green suppliers' evaluation based on AHP & genetic algorithm, in: *2009 International Conference on Signal Processing Systems*, Singapore, 2009, pp. 615–19.

Ghazalli Z, Murata A. Development of an AHP-CBR evaluation system for remanufacturing: End-of-life selection strategy. *International Journal of Sustainable Engineering* 2011;4:2–15.

Ghazilla R, Taha Z, Yusoff S, Rashid S, Sakundarini N. Development of decision support system for fastener selection in product recovery oriented design. *The International Journal of Advanced Manufacturing Technology* 2013;70:1403–13.

Ghorbani M, Arabzad SM, Tavakkoli-Moghaddam R. A multi-objective fuzzy goal programming model for reverse supply chain design. *International Journal of Operational Research* 2014;19:141–53.

Goldberg DE. *Genetic Algorithms in Search, Optimization, and Machine Learning*, Addison-Wesley, Reading, MA, 1989.

Govindan K, Khodaverdi R, Jafarian A. A fuzzy multi criteria approach for measuring sustainability performance of a supplier based on triple bottom line approach. *Journal of Cleaner Production* 2012;47:345–54.

Govindan K, Sarkis J, Palaniappan M. An analytic network process-based multicriteria decision making model for a reverse supply chain. *The International Journal of Advanced Manufacturing Technology* 2013;68:863–80.

Grisi R, Guerra L, Naviglio G. Supplier performance evaluation for green supply chain management, in: *Business Performance Measurement and Management*, Springer Berlin Heidelberg, 2010, pp. 149–63.

Gungor A. Evaluation of connection types in design for disassembly (DFD) using analytic network process. *Computers & Industrial Engineering* 2006;50:35–54.

Gungor A, Gupta SM. Disassembly line in product recovery. *International Journal of Production Research* 2002;40:2569–89.

Gungor A, Gupta SM. Issues in environmentally conscious manufacturing and product recovery: A survey. *Computers & Industrial Engineering* 1999;36: 811–53.

Guo X, Liu S, Wang D, Hou C. An improved multi-objective scatter search approach for solving selective disassembly optimization problem, in: *Proceedings of the 31st Chinese Control Conference*, 2012, pp. 7703–8.

Gupta A, Evans GW. A goal programming model for the operation of closed-loop supply chains. *Engineering Optimization* 2009;41:713–35.

Gupta SM, Isaacs JA. Value analysis of disposal strategies for automobiles. *Computers & Industrial Engineering* 1997;33:325–8.

Gupta SM, Nukala S. A fuzzy AHP-based approach for selecting potential recovery facilities in a closed loop supply chain, in: *Proceedings of the SPIE International Conference on Environmentally Conscious Manufacturing V*, SPIE, Boston, MA, 2005, pp. 58–63.

Gupta SM, Pochampally KK. Evaluation of recycling programs with respect to drivers of public participation: A fuzzy TOPSIS approach, in: *Proceedings of the 2004 Northeast Decision Sciences Institute Conference*, Atlantic City, NJ, 2004, pp. 226–8.

Gupta SM, Taleb K. An algorithm to disassemble multiple product structures with multiple occurrence of parts, in: *Proceedings of the International Seminar on Reuse*, Eindhoven, The Netherlands, 1996, pp. 153–62.

Gupta SM, Veerakamolmal P. A case-based reasoning approach for optimal planning of multi-product/multi-manufacturer disassembly processes. *International Journal of Environmentally Conscious Design and Manufacturing* 2000;9:15–25.

Handfield R, Walton SV, Sroufe R, Melnyk SA. Applying environmental criteria to supplier assessment: A study in the application of the analytical hierarchy process. *European Journal of Operational Research* 2002;141:70–87.

Harraz NA, Galal NM. Design of sustainable end-of-life vehicle recovery network in Egypt. *Ain Shams Engineering Journal* 2011;2:211–19.

Hsu C-W, Hu AH. Application of analytic network process on supplier selection to hazardous substance management in green supply chain management, in: *IEEE International Conference on Industrial Engineering and Engineering Management*, Singapore, 2007, pp. 1362–8.

Hsu C-W, Hu AH. Applying hazardous substance management to supplier selection using analytic network process. *Journal of Cleaner Production* 2009;17:255–64.

Hsu C-W, Hu AH, Chiou C-Y, Chen T-C. Using the FDM and ANP to construct a sustainability balanced scorecard for the semiconductor industry. *Expert Systems with Applications* 2011;38:12891–9.

Hsu CH, Wang F-K, Tzeng G-H. The best vendor selection for conducting the recycled material based on a hybrid MCDM model combining DANP with VIKOR. *Resources, Conservation and Recycling* 2012;66:95–111.

Hula A, Jalali K, Hamza K, Skerlos SJ, Saitou K. Multi-criteria decision-making for optimization of product disassembly under multiple situations. *Environmental Science & Technology* 2003;37:5303–13.

Humphreys P, McCloskey A, McIvor R, Maguire L, Glackin C. Employing dynamic fuzzy membership functions to assess environmental performance in the supplier selection process. *International Journal of Production Research* 2006;44:2379–419.

Humphreys P, McIvor R, Chan F. Using case-based reasoning to evaluate supplier environmental management performance. *Expert Systems with Applications* 2003;25:141–53.

Iakovou E, Moussiopoulos N, Xanthopoulos A, Achillas C, Michailidis N, Chatzipanagioti M, Koroneos C, Bouzakis KD, Kikis V. A methodological framework for end-of-life management of electronic products. *Resources, Conservation and Recycling* 2009;53:329–39.

Ignizio JP. *Goal Programming and Extensions*, Lexington Books, Lexington, 1976.

Ilgin MA, Gupta SM. Environmentally conscious manufacturing and product recovery (ECMPRO): A review of the state of the art. *Journal of Environmental Management* 2010;91: 563–91.

Ilgin MA, Gupta SM. Physical programming: A review of the state of the art. *Studies in Informatics and Control* 2012a;21:349–66.

Ilgin MA, Gupta SM. *Remanufacturing Modeling and Analysis*, CRC Press / Taylor & Francis Group, Boca Raton, FL, 2012b.

Ilgin MA, Gupta SM, Battaia O. Use of MCDM techniques in environmentally conscious manufacturing and product recovery: State of the art. *Journal of Manufacturing Systems* 2015;37:746–58.

Imtanavanich P, Gupta SM. Calculating disassembly yields in a multi-criteria decision-making environment for a disassembly-to-order system, in: K.D. Lawrence, R.K. Klimberg (Eds.) *Applications of Management Science: In Productivity, Finance, and Operations*, Elsevier, Oxford, UK, 2006a, pp. 109–25.

Imtanavanich P, Gupta SM. Evolutionary computation with linear physical programming for solving a disassembly-to-order system, in: *Proceedings of the SPIE International Conference on Environmentally Conscious Manufacturing VI*, Boston, MA, 2006b, pp. 30–41.

Imtanavanich P, Gupta SM. Linear physical programming approach for a disassembly-to-order system under stochastic yields and product's deterioration, in: *Proceedings of the 2006 POMS Meeting*, Boston, MA, 2006c, pp. 004–0213.

Imtanavanich P, Gupta SM. Solving a disassembly-to-order system by using Genetic Algorithm and Weighted Fuzzy Goal Programming, in: *Proceedings of the SPIE International Conference on Environmentally Conscious Manufacturing VI*, Boston, MA, 2006d, pp. 54–56.

Imtanavanich P, Gupta SM. Weighted fuzzy goal programming approach for a disassembly-to-order system, in: *Proceedings of the 2005 POMS-Boston Meeting*, Boston, MA, 2005.

Isaacs JA, Gupta SM. Economic consequences of increasing polymer content for the U.S. automobile recycling infrastructure. *Journal of Industrial Ecology* 1997;1:19–33.

Ishizaka A, Nemery P. *Multi-criteria Decision Analysis: Methods and Software*, John Wiley & Sons, West Sussex, 2013.

Jamshidi R, Fatemi Ghomi SMT, Karimi B. Multi-objective green supply chain optimization with a new hybrid memetic algorithm using the Taguchi method. *Scientia Iranica* 2012;19:1876–86.

Jiang Z, Zhang H, Sutherland JW. Development of multi-criteria decision making model for remanufacturing technology portfolio selection. *Journal of Cleaner Production* 2012;19:1939–45.

Jones DF, Mirrazavi SK, Tamiz M. Multi-objective meta-heuristics: An overview of the current state-of-the-art. *European Journal of Operational Research* 2002;137:1–9.

Jun HB, Cusin M, Kiritsis D, Xirouchakis P. A multi-objective evolutionary algorithm for EOL product recovery optimization: Turbocharger case study. *International Journal of Production Research* 2007;45:4573–94.

Kalayci CB, Gupta SM. Ant colony optimization for sequence-dependent disassembly line balancing problem. *Journal of Manufacturing Technology Management* 2013a;24:413–27.

Kalayci CB, Gupta SM. Artificial bee colony algorithm for solving sequence-dependent disassembly line balancing problem. *Expert Systems with Applications* 2013b;40:7231–41.

Kalayci CB, Gupta SM. Balancing a sequence-dependent disassembly line using simulated annealing algorithm, in: K.D. Lawrence, G. Kleinman (Eds.) *Applications of Management Science*, Emerald, 2013c, pp. 81–103.

Kalayci CB, Gupta SM. A particle swarm optimization algorithm with neighborhood-based mutation for sequence-dependent disassembly line balancing problem. *The International Journal of Advanced Manufacturing Technology* 2013d;69:197–209.

Kalayci CB, Gupta SM. River formation dynamics approach for sequence-dependent disassembly line balancing problem, in: S.M. Gupta (Ed.) *Reverse Supply Chains*, CRC Press, Boca Raton, FL, 2013e, pp. 289–312.

Kalayci CB, Gupta SM. A tabu search algorithm for balancing a sequence-dependent disassembly line. *Production Planning & Control* 2014;25:149–60.

Kannan D, Khodaverdi R, Olfat L, Jafarian A, Diabat A. Integrated fuzzy multi criteria decision making method and multi-objective programming approach for supplier selection and order allocation in a green supply chain. *Journal of Cleaner Production* 2013;47:355–67.

Kannan G, Pokharel S, Sasi Kumar P. A hybrid approach using ISM and fuzzy TOPSIS for the selection of reverse logistics provider. *Resources, Conservation and Recycling* 2009;54:28–36.

Kim J, Hwang Y, Park K. An assessment of the recycling potential of materials based on environmental and economic factors; case study in South Korea. *Journal of Cleaner Production* 2009;17:1264–71.

Kolodner J. An introduction to case-based reasoning. *Artificial Intelligence Review* 1992;6:3–34.

Kongar E, Gupta SM. Disassembly to order system under uncertainty. *Omega* 2006;34:550–61.

Kongar E, Gupta SM. Disassembly-to-order system using linear physical programming, in: *2002 IEEE International Symposium on Electronics* and *the Environment*, San Francisco, CA, USA: IEEE 2002, pp. 312–17.

Kongar E, Gupta SM. A goal programming approach to the remanufacturing supply chain model, in: *Proceedings of the SPIE International Conference on Environmentally Conscious Manufacturing*, Boston, MA, 2000, pp. 167–78.

Kongar E, Gupta SM. Solving the disassembly-to-order problem using linear physical programming. *International Journal of Mathematics in Operational Research* 2009;1:504–31.

Kongar E, Gupta SM, Al-Turki YAY. A fuzzy goal programming approach to disassembly planning in: *The 6th Saudi Engineering Conference*, Dhahran, Saudi Arabia, 2002, pp. 561–79.

Kumar A, Jain V. Supplier selection: A green approach with carbon footprint monitoring, in: *International Conference on Supply Chain Management and Information Systems*, Hong Kong, China, 2010, pp. 1–8.

Kuo TC. Combination of case-based reasoning and analytical hierarchy process for providing intelligent decision support for product recycling strategies. *Expert Systems with Applications* 2010;37:5558–63.

Kuo RJ, Lin YJ. Supplier selection using analytic network process and data envelopment analysis. *International Journal of Production Research* 2012;50:2852–63.

Kuo T-C, Chang S-H, Huang S. Environmentally conscious design by using fuzzy multi-attribute decision-making. *The International Journal of Advanced Manufacturing Technology* 2006;29:419–25.

Kuo Y, Yang T, Huang G-W. The use of grey relational analysis in solving multiple attribute decision-making problems. *Computers & Industrial Engineering* 2008;55:80–93.

Lambert AJD, Gupta SM. *Disassembly Modeling for Assembly, Maintenance, Reuse, and Recycling*, CRC Press, Boca Raton, FL, 2005.

Langella IM. Heuristics for demand-driven disassembly planning. *Computers & Operations Research* 2007;34:552–77.

Lee AHI, Kang H-Y, Hsu C-F, Hung H-C. A green supplier selection model for high-tech industry. *Expert Systems with Applications* 2009;36:7917–27.

Lee SG, Lye SW, Khoo MK. A multi-objective methodology for evaluating product end-of-life options and disassembly. *The International Journal of Advanced Manufacturing Technology* 2001;18:148–56.

Lee T-R, Le TPN, Genovese A, Koh LS. Using FAHP to determine the criteria for partner's selection within a green supply chain: The case of hand tool industry in Taiwan. *Journal of Manufacturing Technology Management* 2012;23:25–55.

Li J, Zhang H-C, Gonzalez MA, Yu S. A multi-objective fuzzy graph approach for modular formulation considering end-of-life issues. *International Journal of Production Research* 2008;46:4011–33.

Li X, Zhao C. Selection of suppliers of vehicle components based on green supply chain, in: *16th International Conference on Industrial Engineering and Engineering Management*, Beijing, 2009, pp. 1588–91.

Lin R-J. Using fuzzy DEMATEL to evaluate the green supply chain management practices. *Journal of Cleaner Production* 2013;40:32–9.

Liu C-H, Huang D-H. Reduction of power consumption and carbon footprints by applying multi-objective optimisation via genetic algorithms. *International Journal of Production Research* 2014;52:337–52.

Lu LYY, Wu CH, Kuo TC. Environmental principles applicable to green supplier evaluation by using multi-objective decision analysis. *International Journal of Production Research* 2007;45:4317–31.

Mahapatara SS, Sharma SK, Parappagoudar MB. A novel multi-criteria decision making approach for selection of reverse manufacturing alternative. *International Journal of Services and Operations Management* 2013;15:176–95.

Mangla S, Madaan J, Sarma PRS, Gupta MP. Multi-objective decision modelling using interpretive structural modelling for green supply chains. *International Journal of Logistics Systems and Management* 2014;17:125–42.

Mareschal B, De Smet Y. Visual PROMETHEE: Developments of the PROMETHEE and GAIA multicriteria decision aid methods, in: *IEEE International Conference on Industrial Engineering and Engineering Management*, Beijing, China, 2009, pp. 1646–9.

Massoud AZ, Gupta SM. Linear physical programming for solving the multi-criteria disassembly-to-order problem under stochastic yields, limited supply, and quantity discount, in: *Proceedings of 2010 Northeast Decision Sciences Institute Conference*, Alexandria, VA, 2010a, pp. 474–9.

Massoud AZ, Gupta SM. Preemptive goal programming for solving the multi-criteria disassembly-to-order problem under stochastic yields, limited supply, and quantity discount, in: *Proceedings of the 2010 Northeast Decision Sciences Institute Conference*, Alexandria,VA, 2010b, pp. 415–20.

McGovern SM, Gupta SM. A balancing method and genetic algorithm for disassembly line balancing. *European Journal of Operational Research* 2007;179:692–708.

McGovern SM, Gupta SM. *The Disassembly Line: Balancing and Modeling*, McGraw-Hill, New York, 2011.

McGovern SM, Gupta SM. Lexicographic goal programming and assessment tools for a combinatorial production problem, in: L.T. Bui, S. Alam (Eds.) *Multi-Objective Optimization in Computational Intelligence: Theory and Practice*, IGI Global, Hershey, PA, 2008, pp. 148–84.

Meade L, Sarkis J. A conceptual model for selecting and evaluating third-party reverse logistics providers. *Supply Chain Management: An International Journal* 2002;7:283–95.

Mehrbod M, Tu N, Miao L, Wenjing D. Interactive fuzzy goal programming for a multi-objective closed-loop logistics network. *Annals of Operations Research* 2012;201:367–81.

Min H. International supplier selection: A multi-attribute utility approach. *International Journal of Physical Distribution & Logistics Management* 1994;24:24–33.

Mirakhorli A. Fuzzy multi-objective optimization for closed loop logistics network design in bread-producing industries. *The International Journal of Advanced Manufacturing Technology* 2014;70:349–62.

Mirhedayatian SM, Azadi M, Saen RF. A novel network data envelopment analysis model for evaluating green supply chain management. *International Journal of Production Economics* 2014;147:544–54.

Nasution PK, Aprilia R, Amalia HM. A goal programming model for the recycling supply chain problem, in: *Proceedings of the 6th IMT-GT Conference on Mathematics, Statistics and its Applications*, Kuala Lumpur Malaysia, 2010, pp. 903–16.

Noci G. Designing a "green" vendor rating systems for the assessment of a supplier's environmental performance. *European Journal of Purchasing & Supply Management* 1997;3:103–14.

Nukala S, Gupta SM. A fuzzy mathematical programming approach for supplier selection in a closed-loop supply chain network, in: *Proceedings of the 2007 POMS-Dallas Meeting*, Dallas, 2007.

Nukala S, Gupta SM. Strategic and tactical planning of a closed-loop supply chain network: A linear physical programming approach, in: *Proceedings of the 2006 POMS Meeting*, Boston, MA, 2006a.

Nukala S, Gupta SM. Supplier selection in a closed-loop supply chain network: An ANP-goal programming based methodology, in: *Proceedings of the SPIE International Conference on Environmentally Conscious Manufacturing VI*, Society of Photo-Optical Instrumentation Engineers, Washington, DC, 2006b, pp. 130–8.

Nurjanni KP, Carvalho MS, da Costa LAAF. Green supply chain design with multi-objective optimization, in: *Proceedings of the 2014 International Conference on Industrial Engineering and Operations Management*, Bali, Indonesia, 2014, pp. 488–97.

Ondemir O, Gupta SM. A multi-criteria decision making model for advanced repair-to-order and disassembly-to-order system. *European Journal of Operational Research* 2014a;233:408–19.

Ondemir O, Gupta SM. Order-driven component and product recovery for sensor-embedded products (SEPS) using linear physical programming, in: *Proceedings of the 41st International Conference on Computers & Industrial Engineering*, Los Angeles, CA, 2011.

Ondemir O, Gupta SM. Quality management in product recovery using the Internet of Things: An optimization approach. *Computers in Industry* 2014b;65:491–504.

Opricovic S, Tzeng G-H. Compromise solution by MCDM methods: A comparative analysis of VIKOR and TOPSIS. *European Journal of Operational Research* 2004;156:445–55.

Ozceylan E, Paksoy T. Fuzzy multi-objective linear programming approach for optimising a closed-loop supply chain network. *International Journal of Production Research* 2013a;51:2443–61.

Ozceylan E, Paksoy T. A mixed integer programming model for a closed-loop supply-chain network. *International Journal of Production Research* 2013b;51:718–34.

Özkır V, Başlıgil H. Multi-objective optimization of closed-loop supply chains in uncertain environment. *Journal of Cleaner Production* 2013;41:114–25.

Paksoy T, Pehlivan NY, Ozceylan E. Fuzzy multi-objective optimization of a green sup-
    ply chain network with risk management that includes environmental hazards.
    Human and Ecological Risk Assessment: An International Journal 2012;18:1120–51.

Pegden CD, Shannon RE, Sadowski RP. *Introduction to Simulation Using SIMAN*,
    McGraw-Hill, 1995.

Pishvaee MS, Razmi J. Environmental supply chain network design using multi-
    objective fuzzy mathematical programming. *Applied Mathematical Modelling*
    2012;36:3433–46.

Pochampally KK, Gupta SM. A linear physical programming approach for design-
    ing a reverse supply chain, in: *Proceedings of the Fifth International Conference on
    Operations and Quantitative Management Seoul*, South Korea, 2004, pp. 261–69.

Pochampally KK, Gupta SM. A multiphase fuzzy logic approach to strategic planning
    of a reverse supply chain network. *IEEE Transactions on Electronics Packaging
    Manufacturing* 2008;31:72–82.

Pochampally KK, Gupta SM. Use of linear physical programming and Bayesian
    updating for design issues in reverse logistics. *International Journal of Production
    Research* 2012;50:1349–59.

Pochampally KK, Gupta SM, Govindan K. Metrics for performance measurement of
    a reverse/closed-loop supply chain. *International Journal of Business Performance
    and Supply Chain Modelling* 2009a;1:8–32.

Pochampally KK, Gupta SM, Kamarthi SV. Identification of potential recovery facili-
    ties for designing a reverse supply chain network using physical programming,
    in: *Proceedings of the SPIE International Conference on Environmentally Conscious
    Manufacturing III*, Providence, RI, 2003, pp. 139–46.

Pochampally KK, Nukala S, Gupta SM. Quantitative decision-making techniques for
    reverse/closed-loop supply chain design, in: S.M. Gupta, A.J.D. Lambert (Eds.)
    Environment Conscious Manufacturing, CRC Press, Boca Raton, FL, 2008.

Pochampally KK, Nukala S, Gupta SM. *Strategic Planning Models for Reverse and
    Closed-Loop Supply Chains*, CRC Press, Boca Raton, FL, 2009b.

Punniyamoorty M, Mathiyalagan P, Lakshmi G. A combined application of struc-
    tural equation modeling (SEM) and analytic hierarchy process (AHP) in sup-
    plier selection. *Benchmarking: An International Journal* 2012;19:70–92.

Ramezani M, Bashiri M, Tavakkoli-Moghaddam R. A new multi-objective stochastic
    model for a forward/reverse logistic network design with responsiveness and
    quality level. *Applied Mathematical Modelling* 2013;37:328–44.

Rao RV. An improved compromise ranking method for evaluation of environmen-
    tally conscious manufacturing programs. *International Journal of Production
    Research* 2009;47:4399–412.

Rao RV, Padmanabhan KK. Selection of best product end-of-life scenario using
    digraph and matrix methods. *Journal of Engineering Design* 2010;21:455–72.

Ravi V. Selection of third-party reverse logistics providers for end-of-life computers
    using TOPSIS-AHP based approach. *International Journal of Logistics Systems and
    Management* 2012;11:24–37.

Ravi V, Shankar R, Tiwari MK. Analyzing alternatives in reverse logistics for end-of-
    life computers: ANP and balanced scorecard approach. *Computers & Industrial
    Engineering* 2005;48:327–56.

Ravi V, Shankar R, Tiwari MK. Selection of a reverse logistics project for end-of-
    life computers: ANP and goal programing approach. *International Journal of
    Production Research* 2008;46:4849–70.

Remery M, Mascle C, Agard B. A new method for evaluating the best product end-of-life strategy during the early design phase. *Journal of Engineering Design* 2012;23:419–41.

Rickli JL, Camelio JA. Multi-objective partial disassembly optimization based on sequence feasibility. *Journal of Manufacturing Systems* 2013;32:281–93.

Roostaee R, Izadikhah M, Lotfi FH. An interactive procedure to solve multi-objective decision-making problem: An improvment to STEM method. *Journal of Applied Mathematics* 2012;2012:1–18.

Saaty TL. *The Analytic Hierarchy Process*, McGraw-Hill, New York, 1980.

Saaty TL. *Decision Making with Dependence and Feedback: The Analytic Network Process*, RWS Publications, Pittsburgh, PA, 1996.

Saen RF. A decision model for selecting third-party reverse logistics providers in the presence of both dual-role factors and imprecise data. *Asia-Pacific Journal of Operational Research* 2011;28:239–54.

Saen RF. A mathematical model for selecting third-party reverse logistics providers. *International Journal of Procurement Management* 2009;2:180–90.

Saen RF. A new model for selecting third-party reverse logistics providers in the presence of multiple dual-role factors. *The International Journal of Advanced Manufacturing Technology* 2010;46:405–10.

Sakundarini N, Taha Z, Abdul-Rashid S, Ghazilla R, Gonzales J. Multi-objective optimization for high recyclability material selection using genetic algorithm. *The International Journal of Advanced Manufacturing Technology* 2013;68:1441–51.

Samanlioglu F. A multi-objective mathematical model for the industrial hazardous waste location-routing problem. *European Journal of Operational Research* 2013;226:332–40.

Samantra C, Sahu NK, Datta S, Mahapatra SS. Decision-making in selecting reverse logistics alternative using interval-valued fuzzy sets combined with VIKOR approach. *International Journal of Services and Operations Management* 2013;14:175–96.

Sangwan KS. Evaluation of manufacturing systems based on environmental aspects using a multi-criteria decision model. *International Journal of Industrial and Systems Engineering* 2013;14:40–57.

Sarkis J. Evaluating environmentally conscious business practices. *European Journal of Operational Research* 1998;107:159–74.

Sarkis J. A methodological framework for evaluating environmentally conscious manufacturing programs. *Computers & Industrial Engineering* 1999;36:793–810.

Sarmiento R, Thomas A. Identifying improvement areas when implementing green initiatives using a multitier AHP approach. *Benchmarking: An International Journal* 2010;17:452–63.

Sasikumar P, Haq AN. Integration of closed loop distribution supply chain network and 3PRLP selection for the case of battery recycling. *International Journal of Production Research* 2011;49:3363–85.

Senthil S, Srirangacharyulu B, Ramesh A. A decision making methodology for the selection of reverse logistics operating channels. *Procedia Engineering* 2012;38:418–28.

Senthil S, Srirangacharyulu B, Ramesh A. A robust hybrid multi-criteria decision making methodology for contractor evaluation and selection in third-party reverse logistics. *Expert Systems with Applications* 2014;41:50–8.

Shaik M, Abdul-Kader W. Green supplier selection generic framework: A multi-attribute utility theory approach. *International Journal of Sustainable Engineering* 2011;4:37–56.

Shaik M, Abdul-Kader W. Performance measurement of reverse logistics enterprise: A comprehensive and integrated approach. *Measuring Business Excellence* 2012;16:23–34.

Shaverdi M, Heshmati MR, Eskandaripour E, Tabar AAA. Developing sustainable SCM evaluation model using fuzzy AHP in publishing industry. *Procedia Computer Science* 2013;17:340–9.

Shen L, Olfat L, Govindan K, Khodaverdi R, Diabat A. A fuzzy multi criteria approach for evaluating green supplier's performance in green supply chain with linguistic preferences. *Resources, Conservation and Recycling* 2013;74:170–9.

Shokohyar S, Mansour S. Simulation-based optimisation of a sustainable recovery network for waste from electrical and electronic equipment (WEEE). *International Journal of Computer Integrated Manufacturing* 2013;26:487–503.

Subramoniam R, Huisingh D, Chinnam RB. Aftermarket remanufacturing strategic planning decision-making framework: Theory and practice. *Journal of Cleaner Production* 2010;18:1575–86.

Subramoniam R, Huisingh D, Chinnam RB, Subramoniam S. Remanufacturing decision-making framework (RDMF): Research validation using the analytical hierarchical process. *Journal of Cleaner Production* 2013;40:212–20.

Taleb KN, Gupta SM. Disassembly of multiple product structures. *Computers & Industrial Engineering* 1997;32:949–61.

Tang Y, Zhou M, Zussman E, Caudill R. Disassembly modeling, planning, and application. *Journal of Manufacturing Systems* 2002;21:200–17.

Teixeira de Almeida A. Multicriteria decision model for outsourcing contracts selection based on utility function and ELECTRE method. *Computers & Operations Research* 2007;34:3569–74.

Thongchattu C, Siripokapirom S. Green supplier selection consensus by neural network, in: *2nd International Conference on Mechanical and Electronics Engineering*, Kyoto, Japan, 2010, pp. 313–16.

Toloie-Eshlaghy A, Homayonfar M. MCDM methodologies and applications: A literature review from 1999 to 2009. *Research Journal of International Studies* 2011;21:86–137.

Tuzkaya G, Gulsun B. Evaluating centralized return centers in a reverse logistics network: An integrated fuzzy multi-criteria decision approach. *International Journal of Environmental Science and Technology* 2008;5:339–52.

Tuzkaya G, Ozgen A, Ozgen D, Tuzkaya UR. Environmental performance evaluation of suppliers: A hybrid fuzzy multi criteria decision approach. *International Journal of Environment Science and Technology* 2009;6:477–90.

Tzeng G-H, Huang JJ. *Multiple Attribute Decision Making: Methods and Applications*, CRC Press, Boca Raton, FL, 2011.

Vadde S, Zeid A, Kamarthi SV. Pricing decisions in a multi-criteria setting for product recovery facilities. *Omega* 2011;39:186–93.

Veerakamolmal P, Gupta SM. A case-based reasoning approach for automating disassembly process planning. *Journal of Intelligent Manufacturing* 2002;13:47–60.

Vinodh S, Mulanjur G, Thiagarajan A. Sustainable concept selection using modified fuzzy TOPSIS: A case study. *International Journal of Sustainable Engineering* 2013;6:109–16.

Vinodh S, Prasanna M, Manoj S. Application of analytical network process for the evaluation of sustainable business practices in an Indian relays manufacturing organization. *Clean Technologies and Environmental Policy* 2012;14:309–17.

Wadhwa S, Madaan J, Chan FTS. Flexible decision modeling of reverse logistics system: A value adding MCDM approach for alternative selection. *Robotics and Computer-Integrated Manufacturing* 2009;25:460–9.

Walther G, Schmid E, Kramer S, Spengler T. Planning and evaluation of sustainable reverse logistics systems, in: *Operations Research Proceedings* 2005, Springer Berlin Heidelberg, 2006, pp. 577–82.

Wang F, Lai X, Shi N. A multi-objective optimization for green supply chain network design. *Decision Support Systems* 2011;51:262–9.

Wang L-C, Tan M-C, Chiang I-F, Chen Y-Y, Lin S-C, Chen ST. Genetic algorithm approach for multi-objective green supply chain design, in: *Proceedings of the 2014 International Conference on Industrial Engineering and Operations Management*, Bali, Indonesia, 2014, pp. 2960–8.

Wang X, Chan HK. A hierarchical fuzzy TOPSIS approach to assess improvement areas when implementing green supply chain initiatives. *International Journal of Production Research* 2013;51:3117–30.

Wang X, Chan HK, Li D. A case study of AHP based model for green product design selection, in: *Proceedings of the EWG-DSS Liverpool-2012 Workshop on Decision Support Systems and Operations Management Trends and Solutions in Industries*, Liverpool, 2012, pp. 1–6.

Wen UP, Chi JM. Developing green supplier selection procedure: A DEA approach, in: *IEEE 17th International Conference on Industrial Engineering and Engineering Management*, 2010, pp. 70–4.

Wittstruck D, Teuteberg F. Integrating the concept of sustainability into the partner selection process: A fuzzy-AHP-TOPSIS approach. *International Journal of Logistics Systems and Management* 2012;12:195–226.

Wu W-W. Choosing knowledge management strategies by using a combined ANP and DEMATEL approach. *Expert Systems with Applications* 2008;35:828–35.

Xanthopoulos A, Iakovou E. On the optimal design of the disassembly and recovery processes. *Waste Management* 2009;29:1702–11.

Xu LD. Case based reasoning. *IEEE Potentials* 1994;13:10–13.

Yang Y, Wu L. Grey entropy method for green supplier selection, in: *Wireless Communications, Networking and Mobile Computing, 2007. WiCom 2007. International Conference on*, Shanghai,China, 2007, pp. 4682–5.

Yeh W-C, Chuang M-C. Using multi-objective genetic algorithm for partner selection in green supply chain problems. *Expert Systems with Applications* 2011;38:4244–53.

Yeh C-H, Xu Y. Sustainable planning of e-waste recycling activities using fuzzy multicriteria decision making. *Journal of Cleaner Production* 2013;52:194–204.

Yu Y, Jin K, Zhang HC, Ling FF, Barnes D. A decision-making model for materials management of end-of-life electronic products. *Journal of Manufacturing Systems* 2000;19:94–107.

Yu-zhong Y, Li-yun W. Extension method for green supplier selection, in: *Wireless Communications, Networking and Mobile Computing, 2008. WiCOM '08. 4th International Conference on*, Dalian, China, 2008, pp. 1–4.

Zareinejad M, Javanmard H. Evaluation and selection of a third-party reverse logistics provider using ANP and IFG-MCDM methodology. *Life Science Journal* 2013;10:350–5.

Zeid I, Gupta SM, Bardasz T. A case-based reasoning approach to planning for disassembly. *Journal of Intelligent Manufacturing* 1997;8:97–106.

Zhang HC, Kuo TC, Lu H, Huang SH. Environmentally conscious design and manufacturing: A state-of-the-art survey. *Journal of Manufacturing Systems* 1997;16: 352–71.

Zhang HC, Li J, Merchant ME. Using fuzzy multi-agent decision-making in environmentally conscious supplier management. *CIRP Annals: Manufacturing Technology* 2003;52:385–8.

Zhou H, Du G, An T. Selection of optimal third-party logistics recycler based on fuzzy DEA, in: *Proceedings of the 2012 International Conference on Automobile and Traffic Science, Materials, Metallurgy Engineering*, Wuhan, China, 2012, pp. 146–51.

Ziout A, Azab A, Altarazi S, ElMaraghy WH. Multi-criteria decision support for sustainability assessment of manufacturing system reuse. *CIRP Journal of Manufacturing Science and Technology* 2013;6:59–69.

2

Techniques Used in the Book

This book illustrates the use of multiple criteria decision making (MCDM) techniques in environmentally conscious manufacturing and product recovery (ECMPRO). In this chapter, we provide an overview of various MCDM techniques. The chapter is organized as follows: Sections 2.2 through 2.14 discuss the basic concepts of many of the MCDM techniques used in this book. Some conclusions are presented in Section 2.15.

2.1 Goal Programming

Goal programming is a popular multiobjective optimization technique. It can be thought of as an extension of linear programming due to its ability to deal with multiple conflicting criteria (Ignizio 1976). Following the determination of a target value for each objective function, the sum of the deviations from target values is minimized. A goal programming study can be conducted by following the steps presented here (Steuer 1986):

- Each objective is conceptualized as a goal.
- The priority and/or weight of each goal is determined.
- Target values of goals are determined.
- The underachievement and overachievement of goals are represented by employing positive and negative deviation variables.
- The decision maker presents his/her desire for the overachievement, underachievement, or satisfaction of the target value exactly for a goal.

There are two common types of goal programming: preemptive and nonpreemptive. The decision maker determines the order of importance of the goals in preemptive goal programming. Then, standard linear programming is used to satisfy the goals based on this order. In nonpreemptive goal programming, first, detrimental deviations are assessed by assigning relative weights to each deviation. Each weight represents the per unit penalty for not meeting a specific goal. Then, these weights are used to convert the goal-programming model into a standard linear-programming model and the total weighted deviation from the goals is minimized using the linear-programming model.

2.2 Fuzzy Logic

Human reasoning modeled in Boolean logic is based on two truth values (i.e., true and false). This is sometimes not suitable to describe the vagueness associated with the human thinking process. Fuzzy logic (FL), first proposed by Zadeh (1965), extends Boolean logic by considering the whole interval between 0 (false) and 1 (true). FL achieves the quantification of the vagueness by using fuzzy set theory. A membership function defined in the interval [0,1] is employed in a fuzzy set to represent the numerical degree of membership of each object belonging to the set. The following expression can be used to define a fuzzy set:

$$\forall y \in Y, \quad \mu_N(y) \in [0,1] \tag{2.1}$$

where:

N is a fuzzy set defined in the universe of quantified linguistic values Y.

y is a generic element of Y.

$\mu_N(y)$ is a membership function associating with each value in Y a real number in the interval [0,1].

A membership function value of 1 represents a perfect match between the generic element y and the fuzzy set N. A generic element with a membership function of 0 does not belong to the fuzzy set. Membership functions can take many different shapes (e.g., triangular, trapezoidal). Graphical depiction of a triangular fuzzy number (TFN) is given in Figure 2.1.

The membership function of a TFN can be defined as follows:

$$\mu_N = \begin{cases} 0 & y < a_1 \\ (x - a_1)/(b_1 - a_1) & a_1 \leq y \leq b_1 \\ (c_1 - x)/(c_1 - b_1) & b_1 \leq y \leq c_1 \\ 0, & y > c_1 \end{cases} \tag{2.2}$$

A fuzzy number is converted into a crisp real number by means of a technique called *defuzzification*. There are several defuzzification techniques.

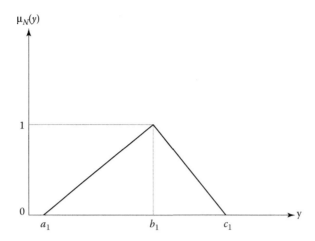

FIGURE 2.1
Triangular fuzzy number.

The center-of-area method, a commonly used defuzzification technique, can be implemented for a TFN ($N = [a_1,b_1,c_1]$) as follows (Tsaur et al. 2002):

$$Q = \frac{(c_1 - a_1) + (b_1 - a_1)}{3} + a_1 \qquad (2.3)$$

The operations defined on TFNs (Chan et al. 2003; Tsaur et al. 2002) can be presented as follows ($F_1 = [a_1,b_1,c_1]$ and $F_2 = [a_2,b_2,c_2]$):
Addition operation:

$$F_1 + F_2 = \left(a_1 + a_2, b_1 + b_2, c_1 + c_2\right) \qquad (2.4)$$

Subtraction operation:

$$F_1 - F_2 = \left(a_1 - c_2, b_1 - b_2, c_1 - a_2\right) \qquad (2.5)$$

Multiplication operation:

$$F_1 \times F_2 = \left(a_1 \times a_2, b_1 \times b_2, c_1 \times c_2\right) \qquad (2.6)$$

Division operation:

$$\frac{F_1}{F_2} = \left(\frac{a_1}{c_2}, \frac{b_1}{b_2}, \frac{c_1}{a_2} \right) \tag{2.7}$$

Another commonly used fuzzy number is the trapezoidal fuzzy number, which is graphically depicted in Figure 2.2. We can write the membership function of a trapezoidal fuzzy number as follows:

$$\mu_T = \begin{cases} 0, & x < a \\ (x-a)/(b-a) & a \le x \le b \\ 1 & b \le x \le c \\ (d-x)/(d-c) & c \le x \le d \\ 0, & x > d \end{cases} \tag{2.8}$$

Defuzzification of a trapezoidal fuzzy number can be carried out using the following formulas under three different cases (Chu and Velásquez 2009):

If the area *abe* is greater than the area *bdfe*, then the defuzzified trapezoidal fuzzy number is calculated using the following formula:

$$Q = a + \frac{1}{2} \cdot \left(2a^2 - 2b^2 + 2bc + 2bd - 2ac - 2ad \right)^{1/2} \tag{2.9}$$

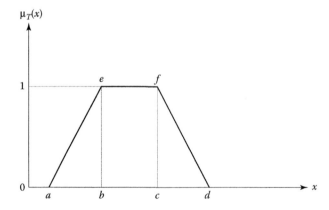

FIGURE 2.2
Trapezoidal fuzzy number.

If the area *acfe* is smaller than the area *cdf*, then the defuzzified trapezoidal fuzzy number is calculated using the following formula:

$$Q = d - \frac{1}{2} \cdot \left(2d^2 - 2c^2 + 2ac + 2bc - 2ad - 2bd\right)^{1/2} \qquad (2.10)$$

Except in these two cases, the following formula is used:

$$Q = \frac{(a+b+c+d)}{4} \qquad (2.11)$$

2.3 Linear Physical Programming

Linear physical programming (Messac et al. 1996) eliminates the weight assignment process encountered in many multiobjective decision making methodologies such as the analytical hierarchy process (AHP), and preferences of decision makers are modeled in a more detailed, quantitative, and qualitative way without using physically meaningless weights (Ilgin and Gupta 2012). The following four classes are used in LPP:

- Smaller-is-better (1S)
- Larger-is-better (2S)
- Value-is-better (3S)
- Range-is-better (4S)

Figure 2.3 presents the class functions. The vertical axis involves the function to be minimized for a criterion (*class function*) while the value of the criterion is given in the horizontal axis. The ideal value of a class function is zero, because a class function with a lower value is better. The value of the *p*th criterion is categorized by using the preference ranges given in the horizontal axis. For instance, we can present these ranges for Class 1S in order of increasing preference as follows:

- $g_p \leq t_{p1}^+$ (Ideal range)
- $t_{p1}^+ \leq g_p \leq t_{p2}^+$ (Desirable range)
- $t_{p2}^+ \leq g_p \leq t_{p3}^+$ (Tolerable range)
- $t_{p3}^+ \leq g_p \leq t_{p4}^+$ (Undesirable range)
- $t_{p4}^+ \leq g_p \leq t_{p5}^+$ (Highly undesirable range)
- $g_p \geq t_{p5}^+$ (Unacceptable range)

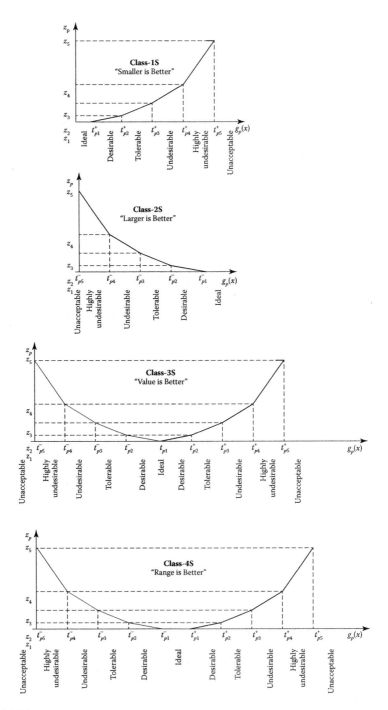

FIGURE 2.3
LPP soft class functions.

Decision makers can use the quantities t^+_{p1} through t^+_{p5} to express their preferences for the pth generic criterion. For instance, a preference vector for the cost criterion can be determined by specifying t^+_{p1} through t^+_{p5} in dollars as (2000, 3000, 4000, 5000, 6000). According to this vector, an alternative having a revenue of $7000 would lie in the unacceptable range; an alternative with a profit of $2400 would lie in the desirable range, and so on.

There are four steps in linear physical programming:

1. The decision maker specifies one of the four classes for each criterion.
2. The decision maker defines the limits of the ranges of differing degrees of desirability (see Figure 2.3) for each criterion.
3. The LPP weight algorithm is used to calculate incremental weights, w^+_{pr} and w^-_{pr} based on the decision maker's preferences. These weights can be considered as the class function slopes. Note that r represents the range intersection.
4. The following expression is employed to calculate a total score (J) for each alternative as a weighted sum of deviations over all ranges and criteria:

$$J = \sum_{p=1}^{N} \sum_{r=2}^{5} \left(w^-_{pr} \cdot d^-_{pr} + w^+_{pr} \cdot d^+_{pr} \right) \tag{2.12}$$

where:
d^-_{pr} is the negative deviation associated with the pth criterion value of the alternative of interest
d^+_{pr} is the positive deviation associated with the pth criterion value of the alternative of interest
N is the total number of criteria

The alternative with the lowest total score value is proposed as the best alternative.

2.4 Data Envelopment Analysis

Data envelopment analysis (DEA) is a linear programming–based technique originally developed by Charnes et al. (1978) for the performance evaluation of multiple decision making units (DMUs) (e.g., schools, hospitals, factories, etc.) with similar inputs and outputs. There are two most commonly used DEA models in the literature: the Charnes, Cooper, and Rho (CCR) model (Charnes et al. 1978) and the Banker, Charnes, and Cooper (BCC) model (Banker et al. 1984). These two models can further be classified into two classes: input-oriented and output-oriented. Input-oriented models minimize the weighted sum of inputs by assuming a fixed level of outputs.. Output-oriented models maximize the weighted sum of outputs by assuming a fixed level of inputs.

2.4.1 CCR Model

The input-oriented CCR model can be written as follows:

$$\max h_o = \frac{\sum_{r=1}^{s} u_r y_{ro}}{\sum_{i=1}^{m} v_i x_{io}}$$

$$\frac{\sum_{r=1}^{s} u_r y_{rj}}{\sum_{i=1}^{m} v_i x_{ij}} \leq 1; \qquad j = 1,\ldots,n$$

$$u_r \geq 0 \qquad\qquad r = 1, 2,\ldots, s \qquad\qquad (2.13)$$

$$v_i \geq 0 \qquad\qquad i = 1, 2,\ldots, m$$

where:
   y_{rj} is the rth known output of the jth DMU
   x_{ij} is the ith known input of the jth DMU
   u_r and v_i are the variable weights determined based on the solution of the problem
  n is the number of DMUs
  m is the number of inputs
  s is the number of outputs

The output-oriented CCR model can be written as follows:

$$\min g_j = \sum_{i=1}^{m} v_i x_{io}$$

$$\sum_{r=1}^{n} u_r y_{ro} = 1$$

$$-\sum_{r=1}^{s} u_r y_{rj} + \sum_{i=1}^{m} v_i x_{ij} \geq 0$$

(2.14)

$$u_r, v_i \geq 0$$

2.4.2 BCC Model

The input-oriented BCC model can be written as follows:

$$\min \theta_o - \varepsilon \left\{ \sum_{i=1}^{m} s_i^- + \sum_{r=1}^{s} s_r^+ \right\}$$

subject to

$$\sum_{j=1}^{n} x_{ij} \lambda_j + s_i^- = \theta_o x_{io} \qquad i = 1, \ldots, m$$

$$\sum_{j=1}^{n} y_{rj} \lambda_j - s_r^+ = y_{ro} \qquad r = 1, \ldots, s$$

$$\sum_{j=1}^{n} \lambda_j = 1$$

$$\lambda_j, s_i^-, s_r^+ \geq 0$$

(2.15)

The output-oriented BCC model can be written as follows:

$$\max \varphi_o + \varepsilon \left\{ \sum_{i=1}^{m} s_i^- + \sum_{r=1}^{s} s_r^+ \right\}$$

subject to

$$\sum_{j=1}^{n} x_{ij} \lambda_j + s_i^- = x_{io} \qquad i = 1, \ldots, m$$

$$\sum_{j=1}^{n} y_{rj} \lambda_j - s_r^+ = \varphi_o y_{ro} \qquad r = 1, \ldots, s$$

$$\sum_{j=1}^{n} \lambda_j = 1$$

$$\lambda_j, s_i^-, s_r^+ \geq 0 \qquad\qquad (2.16)$$

2.5 Analytical Hierarchy Process

AHP (Saaty 1980) is a popular MCDM methodology. It allows decision makers to determine priorities by comparing tangible as well as intangible criteria against each other. Since it has mathematically simple and well-defined steps, researchers have used this technique to solve various problems in different areas, including project management (Mahdi and Alreshaid 2005), health care (Liberatore and Nydick 2008) and military personnel assignment (Korkmaz et al. 2008).

AHP uses a three-level hierarchy to solve an MCDM problem. The first level involves the goal. Decision criteria and decision alternatives are defined in the second and third levels, respectively. AHP is implemented in two steps based on this hierarchy. First, pairwise comparison of independent criteria is achieved using the scale given in Table 2.1 and a suitable technique (e.g., the eigenvalue method) is applied to the comparison matrix to determine the weights of the comparison criteria. Then, pairwise comparison of decision alternatives with respect to each criterion is carried out and the weights of decision alternatives are determined using one of the techniques presented in the first step.

The consistency level of pairwise judgments is measured by calculating a consistency ratio (CR). The following formula is used in this calculation:

$$CR = \frac{(\lambda_{\max} - N)}{(N-1)R} \qquad\qquad (2.17)$$

TABLE 2.1

Scale of Pairwise Judgments

Intensity of Importance	Definition
1	Equally important
3	Moderately more important
5	Strongly important
7	Very strongly more important
9	Extremely more important
2, 4, 6, 8	Intermediate judgment values

TABLE 2.2

Random Index Values

N	1	2	3	4	5	6	7	8	9	10
R	0	0	0.58	0.90	1.12	1.24	0.32	1.41	1.45	1.49

where:

λ_{max} is the principal eigenvalue of the matrix
N is the number of rows (or columns) in the matrix
R is the random index value for each value of N

Table 2.2 presents various R values for values of N ranging from 1 to 10. A CR value of zero represents perfect consistency. The achievement of perfect consistency is very difficult, since humans are prone to error while comparing criteria and alternatives. Hence, a comparison matrix has an acceptable level of consistency if the CR value associated with the matrix is less than 0.1.

2.6 Analytic Network Process

AHP builds a decision hierarchy by assuming that there is no dependence relationship among the criteria, subcriteria, and alternatives. This hierarchical modeling approach contradicts most real-life MCDM problems. The analytic network process (ANP) developed by Saaty (1996) models MCDM problems by using a network approach that allows for dependence within a set of criteria (inner dependence) as well as between sets of criteria (outer dependence); therefore, ANP goes beyond AHP (Ravi et al. 2005).

While AHP assumes a unidirectional hierarchical relationship among the decision levels, ANP allows for a more complex relationship among decision levels and attributes, as it does not require a strict hierarchical structure. The looser network structure in ANP allows the representation of any decision problem, irrespective of what criteria comes first or what comes next. Compared with AHP, ANP requires more calculations and requires a more careful track of the pairwise judgment matrices.

A wide range of ANP applications have been presented in the literature, including the evaluation of third-party reverse logistics providers (Meade and Sarkis 2002), selection of solar thermal power plant investment projects (Aragones-Beltran et al. 2014), and selection of maintenance performance indicators (Van Horenbeek and Pintelon, 2014).

The following steps are followed in an ANP study:

Step 1: Model development (Building a network structure): First, the decision problem is analyzed in a detailed way and its elements are defined as criteria, subcriteria, and alternatives. Then, the clusters associated with each element are constructed. Finally, the ANP network structure is formed by defining the relationships between and within the clusters.

Step 2: Pairwise comparisons: In this step, decision makers are asked to carry out a series of pairwise comparisons where two elements or two clusters are compared according to their relative importance (Saaty 1996). Saaty's nine-point scale is employed in these comparisons.

After completing the pairwise comparisons, a local priority vector is obtained for each comparison matrix by employing mathematical techniques such as the eigenvector method, mean transformation method, or row geometric mean method. This vector shows the relative importance of the elements (or clusters) being compared.

Step 3: Supermatrix formulation: Local priority vectors are entered into the appropriate columns of a partitioned matrix called a *supermatrix*. The submatrices in this matrix represent relationships between and within the clusters.

Step 4: Long-term stable influences of the elements or clusters are obtained by raising the supermatrix to powers. To achieve convergence and obtain the limit supermatrix, the supermatrix is raised to the power of $2k + 1$ where k is an arbitrarily large number. It must be noted that the limit supermatrix can only be obtained if the supermatrix is a column-stochastic matrix (i.e., each column of the matrix sums to unity).

2.7 DEMATEL

DEMATEL, developed by Battelle Memorial Institute (Gabus and Fontela 1972, 1973) is a graph-based methodology used to divide multiple criteria in two groups: cause group and effect group. It involves the following steps:

Step 1: Generate the direct relation matrix. The DEMATEL scale presented in Table 2.3 is used to construct the direct relation matrix A, which is an $n \times n$ matrix obtained by pairwise comparisons. The notation a_{ij} indicates the degree to which the decision maker believes factor i affects factor j. For $i = j$, the diagonal elements are set to zero.

TABLE 2.3

DEMATEL Scale

Linguistic Variable	Influence Score
No influence	0
Very low influence	1
Low influence	2
High influence	3
Very high influence	4

Step 2: Normalize the direct relation matrix. The direct relation matrix is normalized by using the following equations:

$$M = k \times A \qquad (2.18)$$

$$k = \min \left(\frac{1}{\max\limits_{\substack{j=1 \\ 1 \le i \le n}} \sum_{j=1}^{n} |a_{ij}|}, \frac{1}{\max\limits_{\substack{i=1 \\ 1 \le j \le n}} \sum_{i=1}^{n} |a_{ij}|} \right) \qquad (2.19)$$

Step 3: Construct the total relation matrix. The following equations are used to derive the total relation matrix T.

$$T = M(I - M)^{-1} \qquad (2.20)$$

where M is the normalized direct relation matrix and I is the identity matrix.

The sum of the rows and the sum of the columns of the total relation matrix S are computed as D and R using the following equations:

$$D = \left[\sum_{i=1}^{n} t_{ij} \right]_{1 \times n} \qquad (2.21)$$

$$R = \left[\sum_{j=1}^{n} t_{ij} \right]_{1 \times n} \qquad (2.22)$$

The expression $(D + R)$ is called the *prominence,* which indicates the element's degree of influence and being influenced. The expression $(D - R)$ is called the *relation.* If it is positive, the criterion tends to fall under the result category. If it is negative, the criterion tends to fall under the causal category.

Step 4: Set the threshold value and obtain the impact-digraph map. An impact-digraph map can be drawn by considering the elements of the total relations matrix. If all elements are considered, the graph will be too complex. That is why a threshold value is determined by decision makers and the elements whose influence level is greater than the threshold value are included in the impact-digraph map.

2.8 TOPSIS

The technique for order preference by similarity to ideal solution (TOPSIS) method is an MCDM technique that considers the Euclidean distances of decision alternatives to ideal and negative-ideal solutions. According to TOPSIS, the best alternative should have the shortest distance to the ideal solution and the greatest distance from the negative-ideal solution (Hwang and Yoon 1981). The implementation steps of TOPSIS can be presented as follows:

Step 1: A decision matrix providing the rating of each alternative against each criterion is constructed. The decision matrix can be represented as by $A = (a_{ij})_{m \times n}$ for a decision problem involving m alternatives and n criteria. $W = (w_1, w_2, \ldots, w_n)$ is the vector representing the weights of the decision criteria, and the sum of the criteria weights must be equal to 1.

Step 2: A normalized decision matrix is constructed by applying the following formula to each element of the decision matrix:

$$r_{ij} = \frac{a_{ij}}{\sqrt{\sum_{i=1}^{m} a_{ij}^2}}, \quad i = 1, 2, \ldots, m; \quad j = 1, 2, \ldots, n \tag{2.23}$$

Step 3: A weighted normalized decision matrix $[V = (v_{ij})_{m \times n}]$ is constructed using the following formula:

$$v_{ij} = w_j \cdot r_{ij}, \quad i = 1, 2, \ldots, m; \quad j = 1, 2, \ldots, n \tag{2.24}$$

Step 4: Using the elements of the weighted normalized decision matrix, the ideal (A^*) and negative-ideal (A^-) solutions are determined as follows:

$$A^* = \left\{ \max_i v_{ij} \text{ for } i=1,2,\ldots,m \right\} = \{p_1, p_2, \ldots, p_n\} \tag{2.25}$$

$$A^- = \left\{ \min_i v_{ij} \text{ for } i=1,2,\ldots,m \right\} = \{q_1, q_2, \ldots, q_n\} \tag{2.26}$$

Step 5: Euclidean distances of each alternative from the ideal solution and the negative-ideal solution are calculated as follows:

$$E_i^* = \sqrt{\sum_{j=1}^{n} (v_{ij} - v_j^*)^2}, \quad i = 1, 2, \ldots, m \tag{2.27}$$

$$E_i^- = \sqrt{\sum_{j=1}^{n} (v_{ij} - v_j^-)^2}, \quad i = 1, 2, \ldots, m \tag{2.28}$$

Step 6: The following equation is used to calculate the relative closeness of each alternative to the ideal solution:

$$RCC_i = \frac{E_i^-}{E_i^* + E_i^-}, \quad \text{for } i=1,2,\ldots,m \tag{2.29}$$

Step 7: A ranking of decision alternatives is obtained based on the relative closeness values. The best alternative has the highest *RCC* value.

2.9 ELECTRE

ELimination and Choice Expressing Reality (ELECTRE) is an MCDM methodology that builds outranking relations based on pairwise comparisons among decision alternatives for each criterion separately (Sengupta 2016). A decision matrix must be constructed prior to the application of ELECTRE to an MCDM problem. In this matrix, each alternative is given a rating with respect to each criterion. If there are *m* alternatives and *n* criteria, then the decision matrix can be denoted by $X = (x_{ij})_{m \times n}$ while the relative weight vector for criteria can be presented as $W = (w_1, w_2, \ldots, w_n)$. It must be noted that $\sum_{j=1}^{n} w_j = 1$. Following the construction of the decision matrix, the following steps are followed in ELECTRE:

Step 1: The decision matrix is normalized. The following formula is used to calculate an element of the normalized decision matrix R:

$$r_{ij} = \frac{x_{ij}}{\sqrt{\sum_{i=1}^{m} x_{ij}^2}}, \quad i = 1, 2, \ldots, m; \quad j = 1, 2, \ldots, n \tag{2.30}$$

Step 2: The weighted normalized decision matrix $(V = (v_{ij})_{m \times n})$ is constructed as follows:

$$v_{ij} = w_j \cdot r_{ij}, \quad i = 1, 2, \ldots, m; \quad j = 1, 2, \ldots, n \tag{2.31}$$

Step 3: The concordance and discordance sets are determined as follows:

$$C_{ab} = \left\{ j \middle| v_{aj} \geq v_{bj} \right\} \tag{2.32}$$

$$D_{ab} = \left\{ j \middle| v_{aj} < v_{bj} \right\} = J - C_{ab} \tag{2.33}$$

Step 4: The concordance matrix (C) is calculated. The elements of this matrix are the concordance indices, which are calculated using the following equation:

$$c(a,b) = \sum_{j \in C_{ab}} w_j \tag{2.34}$$

Step 5: The disconcordance matrix (D) is calculated. The elements of this matrix are the disconcordance indices, which are calculated using the following equation:

$$d(a,b) = \frac{\max_{j \in D_{ab}} |v_j(a) - v_j(b)|}{\max_{j \in J} |v_j(a) - v_j(b)|} \tag{2.35}$$

Step 6: The concordance dominance matrix (E) is obtained by comparing each element of the concordance matrix against the average index of concordance ($\bar{c}$):

$$E = \begin{cases} e(a,b) = 1 & \text{if } c(a,b) \geq \bar{c} \\ e(a,b) = 0 & \text{if } c(a,b) < \bar{c} \end{cases} \qquad (2.36)$$

where $\bar{c}$ is calculated as follows:

$$\bar{c} = \frac{\displaystyle\sum_{a=1}^{m}\sum_{b=1}^{m} c(a,b)}{m(m-1)} \qquad (2.37)$$

Step 7: The disconcordance dominance matrix (F) is obtained by comparing each element of the disconcordance matrix against the average index of disconcordance ($\bar{d}$):

$$F = \begin{cases} f(a,b) = 1 & \text{if } d(a,b) \geq \bar{d} \\ f(a,b) = 0 & \text{if } d(a,b) < \bar{d} \end{cases} \qquad (2.38)$$

where $\bar{d}$ is calculated as follows:

$$\bar{d} = \frac{\displaystyle\sum_{a=1}^{m}\sum_{b=1}^{m} d(a,b)}{m(m-1)} \qquad (2.39)$$

Step 8: The net superior (c_a) and inferior values (c_b) are calculated as follows:

$$c_a = \sum_{b=1}^{n} c(a,b) - \sum_{b=1}^{n} c(b,a) \qquad (2.40)$$

$$d_a = \sum_{b=1}^{n} d(a,b) - \sum_{b=1}^{n} d(b,a) \qquad (2.41)$$

The variable c_a sums together the number of competitive superiority for all alternatives. That is why larger values of c_a are preferable. On the other hand, d_a sums together the number of inferiority for all alternatives. That is why smaller values of d_a are preferable.

2.10 PROMETHEE

The preference ranking organization method (PROMETHEE) is a popular outranking method developed at the beginning of the 1980s (Brans and Vincke 1985). The working mechanism of this technique can be explained in seven steps:

Step 1: The data matrix of PROMETHEE is constructed by defining alternatives, evaluation criteria, criteria weights, and evaluation values of alternatives with respect to each criterion (see Table 2.4).

Step 2: Preference functions are defined for the criteria (see Table 2.5). There are six different types of preference functions in PROMETHEE.

Step 3: Joint preference functions are developed by considering alternatives in pairs. The joint preference function for alternatives a and b can be constructed as follows:

$$P(a,b) = \begin{cases} 0 & , \ f(a) \le f(b) \\ p[f(a) - f(b)] & , \ f(a) > f(b) \end{cases} \qquad (2.42)$$

Step 4: A preference index is calculated for each alternative pair by using the joint preference functions. The following expression is used to calculate the preference index for alternatives a and b:

TABLE 2.4

Data Matrix of PROMETHEE

		Evaluation Criteria				
		f_1	f_2	f_3	...	f_k
Alternatives	A	$f_1(a)$	$f_2(a)$	$f_3(a)$	...	$f_k(a)$
	B	$f_1(b)$	$f_2(b)$	$f_3(b)$	...	$f_k(b)$
	C	$f_1(c)$	$f_2(c)$	$f_3(c)$	...	$f_k(c)$
	...	...	...	...	...	...
Weights	w_i	w_1	w_2	w_3	...	w_k

TABLE 2.5

Generalized Preference Functions in PROMETHEE

	Function	Parameters
Type I: Usual criterion	$P(d) = \begin{cases} 0 & d \leq 0 \\ 1 & d > 0 \end{cases}$	NA
Type II: Quasi-Criterion (U-Shape)	$P(d) = \begin{cases} 0 & d \leq q \\ 1 & d > q \end{cases}$	q
Type III: (V-Shape)	$P(d) = \begin{cases} 0 & d \leq 0 \\ \dfrac{d}{p} & 0 < d \leq p \\ 1 & d > p \end{cases}$	p
Type IV: (Level-Criterion)	$P(d) = \begin{cases} 0 & d \leq q \\ \dfrac{1}{2} & q < d \leq p \\ 1 & d > p \end{cases}$	p, q
Type V: (Linear Criterion)	$P(d) = \begin{cases} 0 & d \leq p \\ \dfrac{d-q}{p-q} & q < d \leq p \\ 1 & d > p \end{cases}$	p, q
Type VI: (Gaussian Criterion)	$P(d) = \begin{cases} 0 & d \leq 0 \\ 1 - e^{-\frac{d^2}{2s^2}} & d > 0 \end{cases}$	s

Source: Brans et al. 1986.

$$\pi(a,b) = \sum_{i=1}^{k} w_i \cdot P_i(a,b) \qquad (2.43)$$

Step 5: Leaving and entering flows are calculated for each alternative as follows:

$$\phi_a^+ = \sum \pi(a,x) \qquad x=(b,c,d,\ldots) \qquad (2.44)$$

$$\phi_a^- = \sum \pi(x,a) \qquad x=(b,c,d,\ldots) \qquad (2.45)$$

Step 6: PROMETHEE I uses leaving and entering flows to obtain a partial ordering of the alternatives. In PROMETHEE I, alternative *a* is preferred over alternative b if at least one of the following conditions is satisfied:

$$\phi_a^+ > \phi_b^+ \text{ and } \phi_a^- < \phi_b^-$$

$$\phi_a^+ > \phi_b^+ \text{ and } \phi_a^- = \phi_b^- \qquad (2.46)$$

$$\phi_a^+ = \phi_b^+ \text{ and } \phi_a^- < \phi_b^-$$

If the following condition is satisfied, alternatives a and b are said to be *indifferent*:

$$\phi_a^+ = \phi_b^+ \text{ and } \phi_a^- = \phi_b^- \qquad (2.47)$$

If one of the following conditions is satisfied, alternatives a and b cannot be compared:

$$\phi_a^+ > \phi_b^+ \text{ and } \phi_a^- > \phi_b^-$$

$$\phi_a^+ < \phi_b^+ \text{ and } \phi_a^- < \phi_b^- \qquad (2.48)$$

Step 7: PROMETHEE II determines the complete ranking of alternatives by using net flows. The net flow associated with an alternative is calculated as follows:

$$\phi_a^{net} = \phi_a^+ - \phi_a^- \qquad (2.49)$$

Two alternatives are compared using net flows as follows:

- Alternative a is preferred over alternative b if $\phi_a^{net} > \phi_b^{net}$
- Alternatives a and b are indifferent if $\phi_a^{net} = \phi_b^{net}$

2.11 VIKOR

The VlseKriterijumska Optimizacija I Kompromisno Resenje (VIKOR) method (Opricovic and Tzeng 2002, 2004) is an MCDM technique that tries to determine a compromise solution that is a feasible solution closest to the ideal solution. The compromise ranking algorithm of VIKOR has the following steps:

Step 1: Determine the best f_j^* and the worst f_j^- values of all criterion functions, $j = 1, 2, ..., n$. If the jth function represents a benefit, then: $f_j^* = \max_i f_{ij}$ and $f_j^- = \min_i f_{ij}$.

Step 2: Compute the values S_i and R_i, $i = 1, 2, ..., m$, by the relations:

$$S_i = \sum_{j=1}^{n} w_j \cdot \frac{(f_j^* - f_{ij})}{(f_j^* - f_j^-)} \tag{2.50}$$

$$R_i = \max_j \left[w_j \cdot \frac{(f_j^* - f_{ij})}{(f_j^* - f_j^-)} \right] \tag{2.51}$$

where w_j is the weight of the jth criterion.

Step 3: Compute the values Q_i, $i = 1, 2, ..., m$ by the relation:

$$Q_i = v \cdot \frac{(S_i - S^*)}{(S^- - S^*)} + (1 - v) \cdot \frac{(R_i - R^*)}{(R^- - R^*)} \tag{2.52}$$

where:

$S^* = \min_i S_i$, $S^- = \max_i S_i$

$R^* = \min_i R_i$, $R^- = \max_i R_i$

v is introduced as the weight of the strategy of the *majority* of criteria (or, the *maximum group utility*); here, $v = 0.5$

Step 4: Rank the alternatives, sorting by the values S, R, and Q, in descending order. The results are three ranking lists.

Step 5: Propose as a compromise solution the alternative (a') that is ranked the best by the measure Q (minimum) if the following two conditions are satisfied:

- *C1: Acceptable advantage*

$$Q(a'') - Q(a') \geq D(Q)$$

where a'' is the alternative in second position in the ranking list by Q and $D(Q) = 1/(J - 1)$, where J is the number of alternatives.

- *C2: Acceptable stability in decision making*

Alternative a' must also be the best ranked by S and/or R. This compromise solution is stable within a decision making process, which could be *voting by majority rule* (when $v > 0.5$ is needed), or *by consensus* ($v \approx 0.5$), or *with veto* ($v < 0.5$). Here, v is the weight of the decision making strategy *the majority of criteria* (or *the maximum group utility*).

If one of the conditions is not satisfied, then a set of compromise solutions is proposed, which consists of:

1. Alternatives a' and a'' if only condition C2 is not satisfied.
2. Alternatives a', a'', ..., $a(M)$ if condition C1 is not satisfied and $a(M)$ is determined by the relation: $Q(a(M)) - Q(a') < D(Q)$ for maximum M (the positions of these alternatives are *in closeness*). The best alternative, ranked by Q, is the one with the minimum value of Q. The main ranking result is the compromise ranking list of alternatives, and the compromise solution with the *advantage rate*.

2.12 MACBETH

The measuring attractiveness by a categorical based evaluation technique method (MACBETH) is a recently developed technique exploiting the principles of multiattribute value theory (Bana e Costa et al. 2002). In this method, a semantic judgment scale is employed to compare two stimuli at a time based on their difference of attractiveness. These semantic judgments are then converted into a numerical scale by using linear programming. The alternatives are ranked based on their overall attractiveness scores. The following expression is employed for the calculation of attractiveness score:

$$s(a) = \sum_{k=1}^{m} w_k \cdot v_k \left(g_k(a) \right) \tag{2.53}$$

where:

$s(a)$	is the overall attractiveness score for alternative a
w_k	is the weight of criterion k
$g_k(a)$	is the performance of the alternative a regarding criterion g_k
$v_k(g_k(a))$	is the attractiveness of the alternative a on a numerical scale

MACBETH methodology is usually implemented using the M-MACBETH software. This interactive decision making tool helps decision makers in the implementation of MACBETH steps. In addition, it checks the consistency of the judgments and suggests alterations for inconsistent judgments.

2.13 Gray Relational Analysis

Gray relational analysis (GRA) is a part of a larger domain called gray theory. In gray theory, black represents complete lack of information and white represents complete information. That is why a system with incomplete information is called a *gray system*. GRA is used to determine the relationship (similarity) between two data series in a gray system. The algorithm of GRA is stated as follows (Wu and Chen 1999):

Step 1: Generate the referential (x_0) and compared series (x_i) as follows:

$$x_0 = \left(x_0(1), x_0(2), ..., x_0(j), ..., x_0(n) \right) \tag{2.54}$$

$$x_i = \left(x_i(1), x_i(2), ..., x_i(j), ..., x_i(n) \right) \tag{2.55}$$

The compared series x_i can be represented in matrix form:

$$\begin{bmatrix} x_1(1) & x_1(2) & \cdots & x_1(n) \\ x_2(1) & x_2(2) & \cdots & x_2(n) \\ \vdots & \vdots & \ddots & \vdots \\ x_n(1) & x_n(2) & \cdots & x_n(n) \end{bmatrix} \tag{2.56}$$

Step 2: Normalize the data set. The series data can be normalized using one of the following three types: *larger-is-better, smaller-is-better,* and *nominal-is-best.*
For larger-is-better data transformation:

$$x_i^*(k) = \frac{x_i(k) - \min_k x_i(k)}{\max_k x_i(k) - \min_k x_i(k)} \tag{2.57}$$

For smaller-is-better data transformation:

$$x_i^*(k) = \frac{\max_k x_i(k) - x_i(k)}{\max_k x_i(k) - \min_k x_i(k)} \tag{2.58}$$

Step 3: Calculate the gray relational coefficient using the following equation:

$$\xi_i(k) = \frac{\Delta_{min} - \gamma \cdot \Delta_{max}}{\Delta x_i(k) + \gamma \cdot \Delta_{max}} \tag{2.59}$$

where γ is the distinguishing coefficient taking values between 0 and 1. Its value is usually taken as 0.5.

$$\Delta x_i(k) = |x_0(k) - x_i(k)| \tag{2.60}$$

$$\Delta_{min} = \min_i \min_k |x_0(k) - x_i(k)| \tag{2.61}$$

$$\Delta_{max} = \max_i \max_k |x_0(k) - x_i(k)| \tag{2.62}$$

Step 4: Calculate the degree of the gray equation coefficient (r_i). If the weights (w_k) of criteria are determined, the gray relational grade is defined as follows:

$$r_i = \sum_{k=1}^{m} w_k \cdot \xi_i(k) \tag{2.63}$$

The gray relational grade represents the level of correlation between the reference and compared series. If the two series are identical, then the value of the gray relational grade will be equal to 1. According to GRA, if any of the alternatives has a higher gray relational grade than others, it is the most important (or optimal) alternative.

2.14 Simulation

Simulation is a powerful tool used in the analysis of complex processes or systems. It involves the development and analysis of models that have the ability to imitate the behavior of the system being analyzed. After validation, a simulation model can be used mainly for the following three purposes (Pegden et al. 1995):

- Analysis of system behavior
- Development of theories and/or hypotheses based on observed behavior
- Prediction of future behavior

There are some steps to be followed in a successful simulation study. We can list them as follows (Law 2007; Pegden et al. 1995; Banks et al. 2001):

- Define and formulate the problem
- Set the objectives and overall project plan
- Collect input data
- Formulate the model representation
- Translate model representation into modeling software
- Verify (debug) the simulation computer program
- Make pilot runs
  - to validate the model to make sure that it produces reasonable results
  - to determine the length of the initialization period, the length of simulation runs, and the number of replications to be made for each run
- Make production runs
- Analyze output data
- Document and report the results
- Implement the results

As indicated in Law (2007), a simulation study is not necessarily a sequential application of these steps. If there is a need, a simulation analyst can return to a previous step.

Simulation models can be classified into three categories based on the changes in system state:

- Discrete-event simulation models (system state changes at discrete points in time).

- Continuous simulation models (system state changes continuously over time).
- Combined discrete–continuous simulation models (system state changes both continuously over time and at discrete points in time).

A simulation program can be developed either using a programming language (e.g., C, C++, Visual Basic) or a simulation package. The most commonly used two simulation packages are Arena (Kelton et al. 2007) and Extend (Laguna and Marklund 2004). All simulation models presented in this book are discrete-event simulation models developed in Arena.

2.15 Conclusions

In this chapter, an overview of various MCDM techniques used in ECMPRO models presented in different chapters of the book was presented. One or more of the techniques introduced in this chapter are used by each of these models.

References

Aragones-Beltran P, Chaparro-Gonzalez F, Pastor-Ferrando J-P, Pla-Rubio A. An AHP (analytic hierarchy process)/ANP (analytic network process)-based multi-criteria decision approach for the selection of solar-thermal power plant investment projects. *Energy* 2014;66:222–38.

Bana e Costa CA, Correa EC, De Corte J-M, Vansnick J-C. Facilitating bid evaluation in public call for tenders: A socio-technical approach. *Omega* 2002;30:227–42.

Banker RD, Charnes A, Cooper WW. Some models for estimating technical and scale inefficiencies in data envelopment analysis. *Management Science* 1984; 30: 1078–92.

Banks J, Carson JS, Nelson BL, Nicol DM. *Discrete-Event System Simulation*, Prentice Hall, Upper Saddle River, NJ, 2001.

Brans JP, Vincke P. A preference ranking organization method: The PROMETHEE method for MCDM. *Management Science* 1985;31:647–56.

Brans JP, Vincke P., and Mareschal, B. How to select and how to rank projects: The PROMETHEE method. *European Journal of Operational Research* 1986; 24: 228–238.

Chan FT, Chan HK, Chan MH. An integrated fuzzy decision support system for multi-criterion decision making problems. *Journal of Engineering Manufacture* 2003;217:11–27.

Charnes A, Cooper WW, Rhodes E. Measuring the efficiency of decision making units. *European Journal of Operational Research* 1978;2:429–44.

Chu TC, Velásquez A. Evaluating corporate loans via a fuzzy MLMCDM approach, in: *18th World IMACS / MODSIM Congress*, Cairns, Australia, 2009, pp. 1493–99.

Gabus A, Fontela E. *Perceptions of the World Problematique: Communication Procedure, Communicating with Those Bearing Collective Responsibility*, Battelle Geneva Research Center, Geneva, Switzerland, 1973.

Gabus A, Fontela E. *World Problems: An Invitation to Further Thought within the Framework of DEMATEL*, Battelle Geneva Research Center, Geneva, Switzerland, 1972.

Hwang CL, Yoon KS. *Multiple Attribute Decision Making: Methods and Applications*, Springer, Berlin, 1981.

Ignizio JP. *Goal Programming and Extensions*, Lexington Books, Lexington, MA, 1976.

Ilgin MA, Gupta SM. Physical programming: A review of the state of the art. *Studies in Informatics and Control* 2012;21:349–66.

Kelton DW, Sadowski RP, Sadowski DA. *Simulation with Arena*, McGraw-Hill, New York, 2007.

Korkmaz I, Gökçen H, Çetinyokus T. An analytic hierarchy process and two-sided matching based decision support system for military personnel assignment. *Information Sciences* 2008;178:2915–27.

Laguna M, Marklund J. *Business Process Modeling, Simulation and Design*, Prentice Hall, 2004.

Law AM. *Simulation Modelling and Analysis*, McGraw-Hill, New York, 2007.

Liberatore MJ, Nydick RL. The analytic hierarchy process in medical and health care decision making: A literature review. *European Journal of Operational Research* 2008;189:194–207.

Mahdi IM, Alreshaid K. Decision support system for selecting the proper project delivery method using analytical hierarchy process (AHP). *International Journal of Project Management* 2005;23:564–72.

Meade L, Sarkis J. A conceptual model for selecting and evaluating third-party reverse logistics providers. *Supply Chain Management: An International Journal* 2002;7:283–95.

Messac A, Gupta SM, Akbulut B. Linear physical programming: A new approach to multiple objective optimization. *Transactions on Operational Research* 1996;8:39–59.

Opricovic S, Tzeng G-H. Compromise solution by MCDM methods: A comparative analysis of VIKOR and TOPSIS. *European Journal of Operational Research* 2004;156:445–55.

Opricovic S, Tzeng G-H. Multicriteria planning of post-earthquake sustainable reconstruction. *Computer-Aided Civil and Infrastructure Engineering* 2002;17:211–20.

Pegden CD, Shannon RE, Sadowski RP. *Introduction to Simulation Using SIMAN*, McGraw-Hill, New York, 1995.

Ravi V, Shankar R, Tiwari MK. Analyzing alternatives in reverse logistics for end-of-life computers: ANP and balanced scorecard approach. *Computers & Industrial Engineering* 2005;48:327–56.

Saaty TL. *The Analytic Hierarchy Process*, McGraw-Hill, New York, 1980.

Saaty TL. *Decision Making with Dependence and Feedback: The Analytic Network Process*, RWS Publications, Pittsburgh, PA, 1996.

Sengupta, RN. Other decision-making models, in: R.N. Sengupta, A. Gupta, J. Dutta (Eds.) *Decision Sciences: Theory and Practice*, CRC Press, Boca Raton, FL, 2016.

Steuer RE. *Multiple Criteria Optimization: Theory, Computation and Application*, Wiley, New York, 1986.

Tsaur S, Chang T, Yen C. The evaluation of airline service quality by fuzzy MCDM. *Tourism Management* 2002;23:107–15.

Van Horenbeek A, Pintelon L. Development of a maintenance performance measurement framework—using the analytic network process (ANP) for maintenance performance indicator selection. *Omega* 2014;42:33–46.

Zadeh LA. Fuzzy sets. *Information and Control* 1965;8:338–53.

3

Goal Programming

In this chapter, we present a goal programming (GP)-based model. The model addresses an environmentally conscious manufacturing and product recovery (ECMPRO)-related problem within a closed-loop supply chain.

3.1 The Model

A single-period transshipment model is presented in this section to determine the number of units of used products of each type to be selected for recovery, the selection of efficient production facilities, and the optimal number of products (used, new, and recovered) transferred across a closed-loop supply chain. It is assumed that the inventory cost of a used product and a recovered product is 20% of its collection and recovery cost, respectively, and that of a newly produced product is 25% of its production cost. There are three goals in the proposed GP model: maximize the total profit in the closed-loop supply chain (PCSC), maximize the revenue from recycling (RRLC), and minimize the number of disposed items (NDSP).

The minimization of the negative deviations from their respective target values are achieved for the first two goals. The third goal aims at the minimization of the environmental damage. That is why the minimization of the positive deviation from the target value is achieved for this goal. The constraints and the expressions used for cost and revenue calculations are presented in the following sections.

3.1.1 Revenues

3.1.1.1 Reuse Revenue

$$\sum_m \sum_i \{X_{mi} \cdot RES_m\} \tag{3.1}$$

where X_{mi} represents the number of units of product type m picked for product recovery at collection center i and RES_m is the total resale revenue of product m.

3.1.1.2 Recycle Revenue

$$\sum_m \sum_i \left\{ (SPY_{mi} - X_{mi}) \cdot IRC_m \cdot WHT_m \cdot RC_m \cdot RCYL_m \right\} \tag{3.2}$$

where:

SPY_{mi} is the supply of used product m at collection center i
IRC_m is the recycling revenue index of product m
WHT_m is the weight of product m
RC_m is the percentage of recyclable contents by weight in product m
$RCYL_m$ is the total recycling revenue of product m

3.1.1.3 New Product Sale Revenue

$$\sum_m \sum_k \sum_l PRS_m \cdot NNP_{mkl} \tag{3.3}$$

where PRS_m is the selling price of one unit of new product of type m and NNP_{mkl} is the decision variable representing the number of new products of type m transported from production facility k to demand center l.

3.1.2 Costs

3.1.2.1 Collection/Retrieval Cost

$$\sum_i \sum_m UC_i \cdot SPY_{mi} \tag{3.4}$$

where UCi is the cost per product retrieved at collection center i.

3.1.2.2 Processing Cost

This cost is calculated by summing the following three cost components: the disassembly cost of used products, the remanufacturing cost of used products, and the production cost of new products. This can be expressed as follows:

$$\left(DYC \cdot \sum_m \sum_i \sum_r DYT_m \cdot F_{mik} \right)$$

$$+ \sum_m \sum_k \sum_l REC_k \cdot G_{mkl} + \sum_m \sum_k \sum_l NEC_k \cdot NNP_{mkl} \tag{3.5}$$

where:

DYC is the disassembly cost per unit time
DYT_m is the disassembly time for product m

F_{mik} is the decision variable representing the number of used products of type m transported from collection center i to production facility k

REC_k is the cost of remanufacturing at production facility k

G_{mkl} is the decision variable representing the number of used products of type m transported from production facility k to demand center l

NEC_k is the cost of producing one unit of new product at production facility k

3.1.2.3 Inventory Cost

This cost is calculated by summing the following three cost components: the carrying cost of used products inventory at the collection center, the carrying cost of recovered products inventory at the production facility, and the carrying cost of newly manufactured products inventory at the production facility. This can be expressed as follows:

$$\sum_m \sum_i \sum_k (UC_i / 5).F_{mik} + \left(\sum_m \sum_k \sum_l \left\{ (REC_k / 5).G_{mkl} + (NEC_k / 4).NNP_{mkl} \right\} \right)$$

$$(3.6)$$

3.1.2.4 Transportation Cost

This cost is calculated by summing the following two cost components: the cost associated with the transportation of used products from collection centers to the production facility and the cost associated with the transportation of remanufactured and new products from production facilities to demand centers. This can be expressed as follows:

$$TSC_{ik} \cdot \sum_m \sum_i \sum_k F_{mik} + TSR_{kl} \cdot \sum_m \sum_k \sum_l (G_{mkl} + NNP_{mkl}) \qquad (3.7)$$

where TSC_{ik} is the cost to transport one unit from collection center i to production facility k and TSR_{kl} is the cost to transport one unit from production facility k to demand center l.

3.1.2.5 Disposal Cost

Disposal cost is calculated by considering the number of units disposed, the cost of disposing of a product, percentage of recyclable content in the product, and the disposal cost index (a number on a scale between 0 and 10; the smaller the number, the easier or less expensive it is to dispose of the product). This can be expressed as follows:

$$\sum_m \sum_i \left\{ (SPY_{mi} - X_{mi}) \cdot IDIS_m \cdot WHT_m \cdot (1 - RC_m) \right\} \cdot UDC_m \qquad (3.8)$$

where $IDIS_m$ is the disposal cost index of product m ($0 =$ lowest, $10 =$ highest) and UDC_m is the disposal cost of product m.

3.1.3 System Constraints

- The following constraint guarantees that the number of used products selected for remanufacturing at a collection center i is equal to the number of used products sent to all production facilities from that collection center. This can be expressed as follows:

$$\sum_k F_{mik} = X_{mi} \tag{3.9}$$

- The following constraint guarantees that the demand at a demand center is fully satisfied by the new or recovered products sent to that demand center. This can be expressed as follows:

$$\sum_k (G_{mkl} + NNP_{mkl}) = D_{ml} \forall k \tag{3.10}$$

where D_{ml} is the net demand for product type m (remanufactured or new) at demand center l.

- It is assumed that the products in the closed-loop supply chain can only be lost due to common cause variations. To consider those variations at a production facility k, a factor (β_k) is used as follows:

$$\sum_l G_{mkl} = \sum_i F_{mik} \cdot \beta_k \forall k \tag{3.11}$$

- The following constraint guarantees that the total number of used products of type m chosen for recovery at collection center i cannot be greater than the total number of used products suitable for recovery. This can be expressed as follows:

$$X_{mi} \leq SPY_{mi} \cdot (1 - pb_m) \tag{3.12}$$

where pb_m is the probability of breakage of product m.

- The following constraint guarantees that the total number of used products of any type collected at all collection centers (before adjusting for the probability of breakage) is equal to or greater than the net demand. This can be expressed as follows:

$$\sum_m \sum_i SPY_{mi} \geq \sum_m \sum_l D_{ml} \tag{3.13}$$

- The following constraint guarantees that the number of recovered products is not greater than the net demand. This can be expressed as follows:

$$\sum_m \sum_i X_{mi} \le \sum_m \sum_l D_{ml} \qquad (3.14)$$

- The following constraint guarantees that used products' space usage at a production facility cannot be greater than the space available for the used product storage at that facility. This can be expressed as follows:

$$a_1 \cdot \sum_m \sum_i F_{mik} \le SCA_{1k} \cdot U_k \qquad (3.15)$$

where:

a_1 is the space occupied by one unit of used product (square units per product)

SCA_{1k} is the storage capacity of production facility k for used products

U_k is the decision variable signifying the selection of production facility k (1 if selected, 0 if not)

- New and remanufactured products' space usage at a production facility cannot be greater than the space available for new and remanufactured products at that production facility (assuming that the same space is used to store new and remanufactured products):

$$\sum_m \sum_l a_2 \cdot (G_{mkl} + NNP_{mkl}) \le SCA_{2k} \cdot U_k \qquad (3.16)$$

where a_2 is the space occupied by one unit of recovered or new product (square units per product) and SCA_{2k} is the storage capacity of production facility k for new and remanufactured products.

- The following constraint guarantees that used products' space usage at a collection center is not greater than the space available for used products at that collection center.

$$a_1 \cdot \sum_m \sum_k F_{mik} \le SCA_i \qquad (3.17)$$

where SCA_i is the storage capacity of collection center i.

- The ratio of throughput to supply of used products of a production facility is checked. If this ratio is equal to or greater than a preset potential value, the facility can be considered *potential*. This control is carried out only for remanufactured products.

$$\left(\frac{THPT_k}{SPYRE_k}\right) \cdot U_k \geq MINTS \tag{3.18}$$

where:
  $MINTS$ is the minimum throughput per supply
  $THPT_k$ is the throughput (considering only remanufactured products) of production facility k
  $SPYRE_k$ is the supply of used products at production facility k, which is different from SPY_{mi}; these are products that are fit for remanufacturing, after accounting for recycled, disposed, and new products

- Nonnegativity Constraints

$$F_{mik}, G_{mkl}, NNP_{mkl}, X_{mi} \geq 0, \forall i, m, k, l \tag{3.19}$$

$$U_k \in [0.1] \forall k, 0 \text{ if facility } k \text{ not selected, 1 if selected} \tag{3.20}$$

3.2 Numerical Example

The closed-loop supply chain considered in this example involves two collection centers, two production facilities, three demand centers, and two product types. The data for the example we use to implement the GP model is as follows:

$UC_i = 0.01$; $SPY_{11} = 60$; $SPY_{12} = 68$; $SPY_{21} = 70$; $SPY_{22} = 60$; $DYC = 0.06$; $DYT_1 = 8$; $DYT_2 = 12$; $REC_1 = 12$; $REC_2 = 11$; $NEC_1 = 35$; $NEC_2 = 50$; $TSC_{11} = 0.02$; $TSC_{12} = 0.08$; $TSC_{21} = 0.6$; $TSC_{22} = 0.3$; $TSR_{11} = 0.03$; $TSR_{12} = 0.2$; $TSR_{13} = 0.4$; $TSR_{21} = 0.05$; $TSR_{22} = 0.01$; $TSR_{23} = 0.03$; $IDIS_1 = 2$; $IDIS_2 = 6$; $WHT_1 = 0.4$; $WHT_2 = 0.9$; $RC_1 = 0.9$; $RC_2 = 0.9$; $UDC_1 = 0.7$; $UDC_2 = 0.4$; $RES_1 = 35$; $RES_2 = 50$; $RCYL_1 = 10$; $RCYL_2 = 10$; $IRC_1 = 6$; $IRC_2 = 8$; $PRS_1 = 15$; $PRS_2 = 10$; $D_{11} = 40$; $D_{12} = 25$; $D_{13} = 30$; $D_{21} = 15$; $D_{22} = 25$; $D_{23} = 30$; $\beta_1 = 0.7$; $\beta_2 = 0.8$; $pb_1 = 0.4$; $pb_2 = 0.3$; $a_1 = 0.6$; $SCA_{11} = 400$; $SCA_{12} = 500$; $SCA_1 = 250$; $SCA_2 = 300$; $a_2 = 0.9$; $SCA_{21} = 500$; $SCA_{22} = 600$; $MINTS = 0.40$.

When this data is used to solve the GP model, using LINGO (v.4), we obtain $PCSC = 8288$ (target = 9000); $RRLC = 9530$ (target = 10,000); $NDSP = 18$ (target = 15).

The solution also indicated that both the production facilities were chosen for network design. In addition, 70% of the net demand was satisfied by remanufactured products and the remaining 30% by newly manufactured products.

3.3 Other Models

GP has been used extensively by researchers to address various issues associated with environmentally conscious manufacturing and product recovery (ECMPRO)-related problems. For example, Kongar and Gupta (2000) considered environmental, physical, and economic issues and developed a preemptive integer GP model for a disassembly-to-order system. The preemptive GP model proposed by Imtanavanich and Gupta (2006) considered stochastic disassembly yields. The heuristic procedures proposed by Langella (2007) were used to obtain a deterministic equivalent of the problem. An extension of Imtanavanich and Gupta (2006) was proposed by Massoud and Gupta (2010). They considered limited supply and quantity discount besides stochastic yields. The advanced remanufacturing-to-order and disassembly-to-order system analyzed by Ondemir and Gupta (2014) used the life cycle data collected, stored, and delivered by the Internet of things. They developed a lexicographic mixed-integer GP model to optimize this system.

The other disassembly-related issues (e.g., disassembly line balancing and component selection for nondestructive disassembly) were also analyzed using GP. McGovern and Gupta (2008) used lexicographic GP to solve a disassembly line balancing problem with the following: minimization of the resources required for disassembly and maximization of the automation of the process and the quality of the parts or materials recovered. The most desirable components of an end-of-life (EOL) product to be nondestructively disassembled were determined in Xanthopoulos and Iakovou (2009) by using lexicographic GP.

Evaluation of recycling activities is another application area of GP. Gupta and Isaacs (1997) investigated the effect of lightweighting on the dismantler and shredder's profitabilities associated with the EOL vehicle recycling infrastructure of the United States using GP. A GP-based methodology was proposed in Isaacs and Gupta (1997) to analyze the changes to the current U.S. vehicle recycling infrastructure, considering their effects on the dismantler and shredder's profitabilities. Aluminum-intensive vehicle (AIV) processing scenarios were analyzed in Boon et al. (2000, 2003). They used GP for the evaluation of materials streams and processed profitabilities for several different scenarios.

Gupta and Evans (2009) considered multiple products and operations in their nonpreemptive GP model developed for the operational planning of closed-loop supply chains. Harraz and Galal (2011) employed GP to determine the locations for different facilities and the amounts to be allocated to different EOL options. Chaabane et al. (2011) considered carbon emissions, suppliers and subcontractors selection, total logistics costs, technology acquisition, and the choice of transportation modes in their GP-based sustainable supply-chain design methodology.

3.4 Conclusions

In this chapter, GP was used to solve an ECMPRO-related problem. In the model, the design of a closed-loop supply chain was considered. An overview of other models was also presented in the chapter.

References

Boon JE, Isaacs JA, Gupta SM. Economic impact of aluminum-intensive vehicles on the U.S. automotive recycling infrastructure. *Journal of Industrial Ecology* 2000;4: 117–34.

Boon JE, Isaacs JA, Gupta SM. End-of-life infrastructure economics for "clean vehicles" in the United States. *Journal of Industrial Ecology* 2003;7: 25–45.

Chaabane A, Ramudhin A, Paquet M. Designing supply chains with sustainability considerations. *Production Planning & Control* 2011;22: 727–41.

Gupta A, Evans GW. A goal programming model for the operation of closed-loop supply chains. *Engineering Optimization* 2009;41: 713–35.

Gupta SM, Isaacs JA. Value analysis of disposal strategies for automobiles. *Computers & Industrial Engineering* 1997;33: 325–8.

Harraz NA, Galal NM. Design of sustainable end-of-life vehicle recovery network in Egypt. *Ain Shams Engineering Journal* 2011;2: 211–9.

Imtanavanich P, Gupta SM. Calculating disassembly yields in a multi-criteria decision-making environment for a disassembly-to-order system, in: K.D. Lawrence, R.K. Klimberg (Eds.) *Applications of Management Science: In Productivity, Finance, and Operations*, Elsevier, Oxford, UK, 2006, pp. 109–25.

Isaacs JA, Gupta SM. Economic consequences of increasing polymer content for the U.S. automobile recycling infrastructure. *Journal of Industrial Ecology* 1997;1: 19–33.

Kongar E, Gupta SM. A goal programming approach to the remanufacturing supply chain model, in: *Proceedings of the SPIE International Conference on Environmentally Conscious Manufacturing*, Boston, MA, 2000, pp. 167–78.

Langella IM. Heuristics for demand-driven disassembly planning. *Computers & Operations Research* 2007;34: 552–77.

Massoud AZ, Gupta SM. Preemptive goal programming for solving the multi-criteria disassembly-to-order problem under stochastic yields, limited supply, and quantity discount, in: *Proceedings of the 2010 Northeast Decision Sciences Institute Conference*, Alexandria, VA, 2010, pp. 415–20.

McGovern SM, Gupta SM. Lexicographic goal programming and assessment tools for a combinatorial production problem, in: L.T. Bui, S. Alam, (Eds.) *Multi-Objective Optimization in Computational Intelligence: Theory and Practice*, IGI Global, 2008, pp. 148–84.

Ondemir O, Gupta SM. Quality management in product recovery using the Internet of things: An optimization approach. *Computers in Industry* 2014;65: 491–504.

Xanthopoulos A, Iakovou E. On the optimal design of the disassembly and recovery processes. *Waste Management* 2009;29: 1702–11.

4

Fuzzy Goal Programming

In this chapter, we present a fuzzy goal programming-based model. The model considers a scenario within a closed-loop supply chain.

4.1 The Model

In this section, a single-period transshipment model is presented that determines the number of units of used product of each type to be chosen for remanufacturing, identifies efficient production facilities, and achieves the optimal transfer of goods across a closed-loop supply chain, in one continuous phase. The goods include used products transferred from the collection centers to the production facilities, and remanufactured and newly manufactured products transferred from the production facilities to the demand centers.

We assume that the inventory cost of a used product and a remanufactured product is 20% of its collection and remanufacturing cost, respectively, and that of a newly produced product is 25% of its production cost. We consider three goals in our fuzzy goal-programming model:

1. Maximization of the total profit in the closed-loop supply chain (*PCSC*)
2. Maximization of the revenue from recycling (*RRLC*)
3. Minimization of the number of disposed items (*NDSP*)

Minimization of the negative deviation from the respective target values is achieved for the first two goals. The third goal aims at the minimization of environmental damage. That is why minimization of the positive deviation from the target value is achieved for this goal. We can express the membership functions of the three goals as follows:

$$
\mu_1 = \begin{cases} 1 & \text{if} \quad PCSC \geq PCSC_U \\ \dfrac{PCSC - PCSC_L}{PCSC_U - PCSC_L} & \text{if} \quad PCSC_L \leq PCSC \leq PCSC_U \\ 0 & \text{if} \quad PCSC \leq PCSC_L \end{cases}
$$

where $PCSC_U$ is the aspiration level of total profit ($PCSC$) and $PCSC_L$ is the lower tolerance level of $PCSC$.

$$\mu_2 = \begin{cases} 1 & \text{if} & RRLC \geq RRLC_U \\ \dfrac{RRLC - RRLC_L}{RRLC_U - RRLC_L} & \text{if} & RRLC_L \leq RRLC \leq RRLC_U \\ 0 & \text{if} & RRLC \leq RRLC_L \end{cases}$$

where $RRLC_U$ is the aspiration level of recycling revenue ($RRLC$) and $RRLC_L$ is the lower tolerance level of $RRLC$.

$$\mu_3 = \begin{cases} 1 & \text{if} & NDSP \leq NDSP_L \\ \dfrac{NDSP_U - NDSP}{NDSP_U - NDSP_L} & \text{if} & NDSP_L \leq NDSP \leq NDSP_U \\ 0 & \text{if} & NDSP \geq NDSP_U \end{cases}$$

where $NDSP_L$ is the aspiration level of number of disposed items ($NDSP$) and $NDSP_U$ is the upper tolerance level of $NDSP$.

According to the concept of Fibonacci numbers, starting with 1 and 2, the weighting values for μ_1, μ_2, and μ_3 are 0.5, 0.33, and 0.17. The objective function for the weighted fuzzy goal-programming model is as follows.

The cost and revenue criteria and the system constraints considered in the model include:

4.1.1 Revenues

4.1.1.1 Reuse Revenue

$$\sum_m \sum_i \{X_{mi} \cdot RES_m\} \tag{4.1}$$

where X_{mi} represents the number of units of product type m chosen for product recovery at collection center i and RES_m is the total resale revenue of product m.

4.1.1.2 Recycle Revenue

$$\sum_m \sum_i \left\{ (SPY_{mi} - X_{mi}) \cdot IRC_m \cdot WHT_m \cdot RC_m \cdot RCYL_m \right\} \tag{4.2}$$

where:

SPY_{mi} is the supply of used product m at collection center i
IRC_m is the recycling revenue index of product m
WHT_m is the weight of product m
RC_m is the percentage of recyclable contents by weight in product m
$RCYL_m$ is the total recycling revenue of product m

4.1.1.3 New Product Sale Revenue

$$\sum_m \sum_k \sum_l PRS_m \cdot NNP_{mkl} \tag{4.3}$$

where PRS_m is the selling price of one unit of new product of type m and NNP_{mkl} is the decision variable representing the number of new products of type m transported from production facility k to demand center l.

4.1.2 Costs

4.1.2.1 Collection/Retrieval Cost

$$\sum_i \sum_m UC_i \cdot SPY_{mi} \tag{4.4}$$

where UC_i is the cost per product retrieved at collection center i.

4.1.2.2 Processing Cost

This cost is calculated by summing the following three cost components: the disassembly cost of used products, the remanufacturing cost of used products, and the production cost of new products. This can be expressed as follows:

$$\left(DYC \cdot \sum_m \sum_i \sum_r DYT_m \cdot F_{mik} \right) + \sum_m \sum_k \sum_l REC_k \cdot G_{mkl} \tag{4.5}$$

$$+ \sum_m \sum_k \sum_l NEC_k \cdot NNP_{mkl}$$

where:

DYC is the disassembly cost per unit time

DYT_m is the disassembly time for product m

F_{mik} is the decision variable representing the number of used products of type m transported from collection center i to production facility k

REC_k is the cost of remanufacturing at production facility k

G_{mkl} is the decision variable representing the number of used products of type m transported from production facility k to demand center l

NEC_k is the cost of producing one unit of new product at production facility k

4.1.2.3 Inventory Cost

This cost is calculated by summing the following three cost components: the carrying cost of used products inventory at the collection center, the carrying cost of recovered products inventory at the production facility, and the carrying cost of newly manufactured products inventory at the production facility. This can be expressed as follows:

$$\sum_m \sum_i \sum_k (UC_i / 5).F_{mik} +$$

$$\left(\sum_m \sum_k \sum_l \{(REC_k / 5).G_{mkl} + (NEC_k / 4).NNP_{mkl}\} \right) \qquad (4.6)$$

4.1.2.4 Transportation Cost

This cost is calculated by summing the following two cost components: the cost associated with the transportation of used products from collection centers to the production facility and the cost associated with the transportation of remanufactured and new products from the production facilities to the demand centers. This can be expressed as follows:

$$TSC_{ik} \cdot \sum_m \sum_i \sum_k F_{mik} + TSR_{kl} \cdot \sum_m \sum_k \sum_l (G_{mkl} + NNP_{mkl}) \qquad (4.7)$$

where TSC_{ik} is the cost to transport one unit from collection center i to production facility k and TSR_{kl} is the cost to transport one unit from production facility k to demand center l.

4.1.2.5 Disposal Cost

The disposal cost is calculated by considering the number of units disposed, the cost of disposing of a product, the percentage of recyclable content in the product, and the disposal cost index (a number on a scale between 0 and 10; the smaller the number, the easier or less expensive it is to dispose of a product).

$$\sum_m \sum_i \left\{ (SPY_{mi} - X_{mi}) \cdot IDIS_m \cdot WHT_m \cdot (1 - RC_m) \right\} \cdot UDC_m \qquad (4.8)$$

where $IDIS_m$ is the disposal cost index of product m ($0 =$ lowest, $10 =$ highest) and UDC_m is the disposal cost of product m.

4.1.3 System Constraints

- The following constraint guarantees that the number of used products selected for remanufacturing at a collection center i is equal to the number of used products sent to all production facilities from that collection center. This can be expressed as follows:

$$\sum_k F_{mik} = X_{mi} \qquad (4.9)$$

- The following constraint guarantees that the demand at a demand center is fully satisfied by the new or recovered products sent to that demand center. This can be expressed as follows:

$$\sum_k (G_{mkl} + NNP_{mkl}) = D_{ml} \ \forall k \qquad (4.10)$$

where D_{ml} is the net demand for product type m (remanufactured or new) at demand center l.

- It is assumed that the products in the closed-loop supply chain can only be lost due to common cause variations. To consider those variations at a production facility k, a factor (β_k) is used as follows:

$$\sum_l G_{mkl} = \sum_i F_{mik} \cdot \beta_k \ \forall k \qquad (4.11)$$

- The following constraint guarantees that the total number of used products of type m chosen for recovery at collection center i cannot be greater than the total number of used products suitable for recovery. This can be expressed as follows:

$$X_{mi} \leq SPY_{mi} \cdot (1 - pb_m) \qquad (4.12)$$

where pb_m is the probability of breakage of product m.

- The following constraint guarantees that the total number of used products of any type collected at all collection centers (before adjusting for the probability of breakage) is equal to or greater than the net demand. This can be expressed as follows:

$$\sum_m \sum_i SPY_{mi} \geq \sum_m \sum_l D_{ml} \qquad (4.13)$$

- The following constraint guarantees that the number of recovered products is not greater than the net demand. This can be expressed as follows:

$$\sum_m \sum_i X_{mi} \leq \sum_m \sum_l D_{ml} \qquad (4.14)$$

- The following constraint guarantees that used products' space usage at a production facility cannot be greater than the space available for the used product storage at that facility. This can be expressed as follows:

$$a_1 \cdot \sum_m \sum_i F_{mik} \leq SCA_{1k} \cdot U_k \qquad (4.15)$$

where:

a_1 is the space occupied by one unit of used product (square units per product)

SCA_{1k} is the storage capacity of production facility k for used products

U_k is the decision variable signifying the selection of the production facility k (1 if selected, 0 if not)

- New and remanufactured products' space usage at a production facility cannot be greater than the space available for new and remanufactured products at that production facility (assuming that the same space is used to store new and remanufactured products):

$$\sum_m \sum_l a_2 \cdot (G_{mkl} + NNP_{mkl}) \leq SCA_{2k} \cdot U_k \qquad (4.16)$$

where a_2 is the space occupied by one unit of recovered or new product (square units per product) and SCA_{2k} is the storage capacity of production facility k for new and remanufactured products.

- The following constraint guarantees that used products' space usage at a collection center is not greater than the space available for used products at that collection center.

$$a_1 \cdot \sum_m \sum_k F_{mik} \leq SCA_i \tag{4.17}$$

where SCA_i is the storage capacity of collection center i.

- The ratio of throughput to supply of used products of a production facility is checked. If this ratio is equal to or greater than a preset potential value, the facility can be considered *potential*. This control is carried out only for remanufactured products.

$$(THPT_k / SPYRE_k) \cdot U_k \geq MINTS \tag{4.18}$$

where:
$MINTS$ is the minimum throughput per supply
$THPT_k$ is the throughput (considering only remanufactured products) of production facility k
$SPYRE_k$ is the supply of used products at production facility k, which is different from SPY_{mi}; these are products that are fit for remanufacturing after accounting for recycled, disposed, and new products

- Nonnegativity Constraints

$$F_{mik}, G_{mkl}, NNP_{mkl}, X_{mi} \geq 0, \forall\, i, m, k, l \tag{4.19}$$

$$U_k \in [0,1] \forall\, k,\ 0 \text{ if facility } k \text{ not selected, } 1 \text{ if selected} \tag{4.20}$$

4.2 Numerical Example

We consider a closed-loop supply chain with three collection centers, two production facilities from which to choose, two demand centers to be served, and three types of products from which to choose. The data for the example we use to implement the fuzzy goal-programming model is as follows:

$UC_i = 0.02$; $SPY_{11} = 60$; $SPY_{12} = 40$; $SPY_{13} = 20$; $SPY_{21} = 30$; $SPY_{22} = 35$; $SPY_{23} = 20$; $SPY_{31} = 25$; $SPY_{32} = 35$; $SPY_{33} = 30$; $DYC = 0.05$; $DYT_1 = 10$; $DYT_2 = 12$; $DYT_3 = 9$; $REC_1 = 13$; $REC_2 = 10$; $NEC_1 = 60$; $NEC_2 = 45$; $TSC_{11} = 0.01$; $TSC_{12} = 0.09$; $TSC_{21} = 0.5$; $TSC_{22} = 0.1$; $TSC_{31} = 0.02$; $TSC_{32} = 0.04$; $TSR_{11} = 0.04$; $TSR_{12} = 0.03$; $TSR_{21} = 0.09$; $TSR_{22} = 0.05$; $IDIS_1 = 4$; $IDIS_2 = 6$; $IDIS_3 = 5$; $WHT_1 = 0.8$; $WHT_2 = 1.0$; $WHT_3 = 0.9$; $RC_1 = 0.5$; $RC_2 = 0.6$; $RC_3 = 0.75$; $UDC_1 = 0.2$; $UDC_2 = 0.5$; $UDC_3 = 0.3$; $RES_1 = 30$; $RES_2 = 40$; $RES_3 = 45$; $RCYL_1 = 1.5$; $RCYL_2 = 2$; $RCYL_3 = 2.5$; $IRC_1 = 7$; $IRC_2 = 4$; $IRC_3 = 5$; $PRS_1 = 65$; $PRS_2 = 53$; $PRS_3 = 60$; $D_{11} = 20$; $D_{12} = 15$; $D_{21} = 16$; $D_{22} = 22$; $D_{31} = 25$; $D_{32} = 20$; $\beta_1 = 0.4$; $\beta_2 = 0.6$; $pb_1 = 0.2$; $pb_2 = 0.4$; $pb_3 = 0.3$; $a_1 = 0.7$; $SCA_{11} = 400$; $SCA_{12} = 400$; $SCA_1 = 150$; $SCA_2 = 150$; $SCA_3 = 150$; $a_2 = 0.7$; $SCA_{21} = 500$; $SCA_{22} = 500$; $PCSC_U = 3500$; $PCSC_L = 2000$; $RRLC_U = 1000$; $RRLC_L = 800$; $NDSP_L = 60$; $NDSP_U = 100$; $MINTS = 0.25$.

When this data is used to solve the fuzzy goal-programming model, using LINGO (v.4), we obtain $PCSC = 3500$; $RRLC = 1000$; $NDSP = 77$; $\mu_1 = 1$; $\mu_2 = 1$; $\mu_3 = 0.59$. It should be noted that the achievements of $PCSC$ and $RRLC$ (see μ_1 and μ_2) are at their maximum, while that of $NDSP$ is not (see μ_3). In addition, both production facilities were chosen for the network design. Also, 86% of the net demand was satisfied by remanufactured products and the remaining 14% by newly manufactured products.

4.3 Other Models

Fuzzy goal programming has been used by many researchers to address various issues associated with environmentally conscious manufacturing and product recovery (ECMPRO) related problems. For example, Kongar et al. (2002) and Kongar and Gupta (2006) used fuzzy goal programming to solve the disassembly-to-order problem. Mehrbod et al. (2012) first developed a multiobjective mixed-integer nonlinear programming formulation for a closed-loop supply chain. Then, this model was solved using interactive fuzzy goal programming (IFGP), which has the ability to address the imprecise nature of decision makers' aspiration levels for goals. Ghorbani et al. (2014) developed a fuzzy goal-programming model for the design of a reverse logistics network. Imtanavanich and Gupta (2005) employed weighted fuzzy goal programming to solve the multiperiod disassembly-to-order (DTO) problem. Imtanavanich and Gupta (2006) integrated genetic algorithms with weighted fuzzy goal programming to solve a similar DTO problem.

4.4 Conclusions

In this chapter, fuzzy goal programming was used to solve an ECMPRO-related problem. In the model, the design of a closed-loop supply chain was considered. An overview of other models was also presented in the chapter.

References

Ghorbani M, Arabzad SM, Tavakkoli-Moghaddam R. A multi-objective fuzzy goal programming model for reverse supply chain design. *International Journal of Operational Research* 2014;19: 141–53.

Imtanavanich P, Gupta SM, Weighted fuzzy goal programming approach for a disassembly-to-order system, in: *Proceedings of the 2005 POMS-Boston Meeting*, Boston, MA, Productions and Operations Management Society, Miami, FL, 2005.

Imtanavanich P, Gupta SM, Solving a disassembly-to-order system by using genetic algorithm and weighted fuzzy goal programming, in: *Proceedings of the SPIE International Conference on Environmentally Conscious Manufacturing VI*, Boston, MA, International Society for Optics and Photonics, Bellinghan, WA, 2006, pp. 54–6.

Kongar E, Gupta SM. Disassembly to order system under uncertainty. *Omega* 2006;34: 550–61.

Kongar E, Gupta SM, Al-Turki YAY, A fuzzy goal programming approach to disassembly planning, in: *The 6th Saudi Engineering Conference*, Dhahran, Saudi Arabia, 2002, pp. 561–79.

Mehrbod M, Tu N, Miao L, Wenjing D. Interactive fuzzy goal programming for a multi-objective closed-loop logistics network. *Annals of Operations Research* 2012;201: 367–81.

5

Linear Physical Programming

Linear physical programming (LPP) eliminates the subjective weight assignment process of weight-based multiple criteria decision making (MCDM) techniques by allowing decision makers to express their preferences in a physically meaningful way. In this chapter, we present an LPP-based model that focuses on the design of a closed-loop supply chain.

5.1 The Model

LPP can be used to design a generic reverse supply chain involving collection centers, remanufacturing facilities, and demand centers. Following the representation of the problem according to LPP principles, optimal transportation quantities for used and remanufactured products are determined. Modeling details are presented below.

5.1.1 Model Formulation

5.1.1.1 Class-1S Criteria (Smaller-Is-Better)

Total transportation cost per period (s_1) can be presented as follows:

$$s_1 = \sum_x \sum_y UTC_{xy} \cdot TS_{xy} + \sum_y \sum_z TRP_{yz} \cdot NR_{yz} \qquad (5.1)$$

where:
UTC_{xy} is the cost of transporting one unit from collection center x to remanufacturing facility y

TS_{xy} is the number of products to be transported from collection center x to remanufacturing facility y

TRP_{yz} is the cost of transporting one unit from remanufacturing facility y to demand center z

NR_{yz} is the number of products to be transported from remanufacturing facility y to demand center z

Total remanufacturing cost per period (s_2) *can be presented as follows:*

$$s_2 = \sum_y \sum_z URC_y \cdot NR_{yz} \qquad (5.2)$$

where:

URC_y is the remanufacturing cost of each product at remanufacturing facility y

NR_{yz} is the number of products to be transported from remanufacturing facility y to demand center z

Total inventory cost per period (s_3) *can be presented as follows:*

$$s_3 = \sum_x \sum_y (RC_x/4) \cdot TS_{xy} + \sum_y \sum_z (URC_y/4) \cdot NR_{yz} \qquad (5.3)$$

where:

RC_x is the retrieval cost of each product at collection center x

TS_{xy} is the number of products to be transported from collection center x to remanufacturing facility y

URC_y is the remanufacturing cost of each product at remanufacturing facility y

NR_{yz} is the number of products to be transported from remanufacturing facility y to demand center z

5.1.1.2 Class-1H Criteria

Total retrieval cost per period (h_1) *can be presented as follows:*

$$h_1 = \sum_x \sum_y RC_x \cdot TS_{xy} \qquad (5.4)$$

where:

RC_x is the retrieval cost of each product at collection center x

TS_{xy} is the number of products to be transported from collection center x to remanufacturing facility y

5.1.1.3 Goal Constraints

$$h_1 \leq R_{max} \qquad (5.5)$$

where R_{max} is the maximum allowed retrieval cost value

$$g_p - d_{pr}^+ \leq t_{p(r-1)}^+ \tag{5.6}$$

$$g_p \leq t_{p5}^+ \tag{5.7}$$

$$d_{pr}^+ \geq 0 \tag{5.8}$$

5.1.1.4 System Constraints

$$\sum_y NR_{yz} = D_z \ \forall z \tag{5.9}$$

where D_z is the total demand for remanufactured products
According to the constraint, demand must be satisfied at each demand center.

$$\sum_x TS_{xy} = C\_Cap_x \ \forall y \tag{5.10}$$

where: C_Cap_x is the supply at collection center x
According to the constraint, each collection center must transport all the used products it received.

$$\sum_z NR_{yz} = \sum_x TS_{xy} \ \forall y \tag{5.11}$$

According to the constraint, the number of used products must be equal to the number of remanufactured products.

$$\sum_x sp \cdot TS_{xy} \leq R\_Cap_y ; \forall y \tag{5.12}$$

where:

sp is the space occupied by one unit of remanufactured or new product (square units per product)

TS_{xy} is the number of products to be transported from collection center x to remanufacturing facility y

R_Cap_y is the storage capacity of remanufacturing facility y for used products

According to the constraint, the space occupied by used products is at most the storage capacity of the remanufacturing facility.

$$TS_{xy} \geq 0; \forall x, y \tag{5.13}$$

$$NR_{yz} \geq 0; \forall y, z \tag{5.14}$$

The decision maker can include additional constraints based on his/her preferences.

5.2 Numerical Example

In this example, we consider a reverse supply chain with two collection centers, three remanufacturing facilities, and two demand centers. The data for the example is as follows:

$RC_1 = 0.4; RC_2 = 0.6; UTC_{1A} = 0.01; UTC_{1B} = 0.3; UTC_{1C} = 0.3; UTC_{2A} = 0.5;$
$UTC_{2B} = 1; UTC_{2C} = 0.5; TRP_{A1} = 2; TRP_{A2} = 5; TRP_{B1} = 0.05; TRP_{B2} = 2;$
$TRP_{C1} = 0.5; \quad TRP_{C2} = 0.5; \quad URC_A = 0.5; \quad URC_B = 0.2; \quad URC_C = 0.3;$
$D_1 = 100; D_2 = 170; C\_Cap_1 = 165; C\_Cap_2 = 200; sp = 1; R\_Cap_A = 500;$
and $R\_Cap_B = R\_Cap_C = 400$. Table 5.1 presents the target values for each soft criterion while the incremental weights obtained by the LPP weight algorithm (Messac et al. 1996) are shown in Table 5.2.

The use of LINGO (v.11) produced the following optimal solution for the model:

TABLE 5.1

Preference Table (in Hundreds)

Criteria	t_{p1}^+	t_{p2}^+	t_{p3}^+	t_{p4}^+	t_{p5}^+
s_1	200	300	400	500	600
s_2	100	150	200	250	300
s_3	90	120	170	200	250

TABLE 5.2

Output of LPP Weight Algorithm

Criteria	$\Delta\omega_{p2}^+$	$\Delta\omega_{p3}^+$	$\Delta\omega_{p4}^+$	$\Delta\omega_{p5}^+$
s_1	0.093914	0.112697	0.247934	0.545455
s_2	0.093914	0.112697	0.247934	0.545455
s_3	0.156524	0.050088	0.550964	0.242424

$TS_{1A} = 0$; $TS_{2A} = 0$; $TS_{1B} = 165$; $TS_{2B} = 105$; $TS_{1C} = 0$; $TS_{2C} = 95$; $NR_{A1} = 0$; $NR_{A2} = 0$; $NR_{B1} = 100$; $N_{B2} = 170$; $NR_{C1} = 0$; and $NR_{C2} = 95$. This solution indicates that the number of products transported from Collection center 1 to Remanufacturing facility B is 165; from Collection center 2 to Remanufacturing facility B is 105; from Collection center 2 to Remanufacturing facility C is 95; from Remanufacturing facility B to Demand center 1 is 100; from Remanufacturing facility B to Demand center 2 is 170; and from Remanufacturing facility C to Demand center 2 is 95. Note that the model did not choose Remanufacturing facility A.

5.3 Other Models

The use of LPP in the design of reverse and closed-loop supply-chain networks is an active research area. Pochampally et al. (2003) used LPP to identify potential facilities from a set of candidate recovery facilities operating in a region. The criteria considered included quality of products at recovery facility, ratio of throughput to supply of used products, multiplication of throughput by disassembly time, and customer service rating of the recovery facility. Pochampally and Gupta (2004) proposed a three-phase methodology for the design of a reverse supply chain. The first phase involved the determination of the economical products to be reprocessed. Potential recovery facilities were determined in the second phase. Finally, the third phase determined the right mix and quantities of products to be transported in the reverse supply chain. Nukala and Gupta (2006) developed a model for the strategic and tactical planning of a closed-loop supply chain. This model can support decision makers on the following issues: the most economical used product to reprocess, efficient production facilities, and the right mix and quantity of goods to be transported across the supply chain. Pochampally et al. (2008), Pochampally et al. (2009b), and Ilgin and Gupta (2012) presented similar models.

Pochampally et al. (2009a) first identified the metrics for the performance evaluation of reverse and closed-loop supply chains. Then, they integrated quality function deployment (QFD) and LPP to determine the "satisfaction level" of a reverse/closed-loop supply chain based on the metrics determined previously.

Pochampally and Gupta (2012) used LPP to develop a collection center selection methodology. They considered the following criteria: sigma level, per capita income of people in residential area, utilization of incentives from local government, distance from residential area, distance from highway, incentives from local government, space cost, and labor cost.

The use of LPP in the analysis of disassembly-to-order (DTO) systems is another active research area. Kongar and Gupta (2002) developed an LPP model to determine the number of items to be disassembled for remanufacturing, recycling, storage, and disposal. They considered the following criteria: average customer satisfaction, average quality achievement, resale revenue, recycling revenue, total profit, number of recycled items, average environmental damage, average environmental benefit, and number of disposed items. The LPP-based solution methodology developed by Kongar and Gupta (2009) for a DTO problem considered several goals including number of disposed items, total profit, number of recycled items, environmental damage, and customer satisfaction. Imtanavanich and Gupta (2006b) employed LPP to solve a DTO problem under stochastic yields and product deterioration. Imtanavanich and Gupta (2006a) integrated genetic algorithms and LPP. They calculated the fitness value of GA by using an LPP model for a DTO system. Massoud and Gupta (2010) developed an LPP-based methodology for a multiperiod DTO problem. This methodology considered several objectives including maximization of profit, minimization of procurement cost, minimization of purchase cost, and minimization of disposal cost. Ondemir and Gupta (2011) considered a DTO system in which sensors and radio-frequency identification (RFID) tags were used to collect, store, and deliver the life cycle data associated with (end-of-life) EOL products. They proposed an LPP model to determine disassembly, refurbishment, disposal, recycling, and storage plans. Ondemir and Gupta (2014) considered sensor embedded EOL products and performed the multicriteria optimization of an advanced repair-to-order and disassembly-to-order system.

5.4 Conclusions

In this chapter, LPP was used to solve an environmentally conscious manufacturing and product recovery (ECMPRO)-related problem. In the model, design of a closed-loop supply chain was considered. An overview of other models was also presented in the chapter.

References

Ilgin MA, Gupta SM. *Remanufacturing Modeling and Analysis*. CRC Press/Taylor & Francis, Boca Raton, FL, 2012.

Imtanavanich P, Gupta SM. Evolutionary computation with linear physical programming for solving a disassembly-to-order system, in: *Proceedings of the SPIE International Conference on Environmentally Conscious Manufacturing VI*, Boston, MA, International Society for Optics and Photonics, Bellington, WA, 2006a, pp. 30–41.

Imtanavanich P, Gupta SM. Linear physical programming approach for a disassembly-to-order system under stochastic yields and product's deterioration, in: *Proceedings of the 2006 POMS Meeting*, Boston, MA, Production and Operations Management Society, Miami, FL, 2006b, pp. 004–0213.

Kongar E, Gupta SM. Disassembly-to-order system using linear physical programming, in: *2002 IEEE International Symposium on Electronics and the Environment*, Institute of Electrical and Electronics Engineers, Piscataway, NJ, 2002, pp. 312–7.

Kongar E, Gupta SM. Solving the disassembly-to-order problem using linear physical programming. *International Journal of Mathematics in Operational Research* 2009;1: 504–31.

Lambert, A.J.D., Gupta, S.M., 2005. *Disassembly Modeling for Assembly, Maintenance, Reuse, and Recycling*. CRC Press, Boca Raton, FL.

Massoud AZ, Gupta SM. Linear physical programming for solving the multi-criteria disassembly-to-order problem under stochastic yields, limited supply, and quantity discount, in: *Proceedings of 2010 Northeast Decision Sciences Institute Conference*, Alexandria, VA, Northeast Decision Sciences Institute, Boston, MA, 2010, pp. 474–9.

Messac A, Gupta SM, Akbulut B. Linear physical programming: A new approach to multiple objective optimization. *Transactions on Operational Research* 1996;8: 39–59.

Nukala S, Gupta SM. Strategic and tactical planning of a closed-loop supply chain network: A linear physical programming approach, in: *Proceedings of the 2006 POMS Meeting*, Boston, MA, Production and Operations Management Society, Miami, FL, 2006, pp. 004–0210.

Ondemir O, Gupta SM. Order-driven component and product recovery for sensor-embedded products (SEPS) using linear physical programming, in: *Proceedings of the 41st International Conference on Computers & Industrial Engineering*, Los Angeles, CA, Computers and Industrial Engineering, California, USA, 2011.

Ondemir O, Gupta SM. A multi-criteria decision making model for advanced repair-to-order and disassembly-to-order system. *European Journal of Operational Research* 2014;233: 408–19.

Pochampally KK, Gupta SM. A linear physical programming approach for designing a reverse supply chain, in: *Proceedings of the Fifth International Conference on Operations and Quantitative Management*, Seoul, South Korea, Operations and Quantitative Management, Texas, USA, 2004, pp. 261–9.

Pochampally KK, Gupta SM. Use of linear physical programming and Bayesian updating for design issues in reverse logistics. *International Journal of Production Research* 2012;50: 1349–59.

Pochampally KK, Gupta SM, Govindan K. Metrics for performance measurement of a reverse/closed-loop supply chain. *International Journal of Business Performance and Supply Chain Modelling* 2009a;1: 8–32.

Pochampally KK, Gupta SM, Kamarthi SV. Identification of potential recovery facili-
    ties for designing a reverse supply chain network using physical programming,
    in: *Proceedings of the SPIE International Conference on Environmentally Conscious
    Manufacturing III*, Providence, RI, 2003, pp. 139–46.
Pochampally KK, Nukala S, Gupta SM. Quantitative decision-making techniques for
    reverse/closed-loop supply chain design, in: S.M. Gupta, A.J.D. Lambert (Eds.)
    Environment Conscious Manufacturing, CRC Press, Boca Raton, FL, 2008.
Pochampally KK, Nukala S, Gupta SM. *Strategic Planning Models for Reverse and
    Closed-Loop Supply Chains*, CRC Press, Boca Raton, FL, 2009b.
Tang Y, Zhou M, Zussman E, Caudill R. Disassembly modeling, planning, and appli-
    cation. *Journal of Manufacturing Systems* 2002;21: 200–17.

6

Data Envelopment Analysis

In this chapter, a data envelopment analysis (DEA)-based model that focuses on the environmental evaluation of suppliers is presented.

6.1 The Model

The modeling details are presented below.

The following inputs are considered in the DEA model:

- *Price*: The price ($/lb) offered by a supplier for the material considered in this model.
- *Carbon footprint*: Total carbon footprint of a supplier represented in megatons of CO_2. This measure is calculated by considering the key emission sources of a supplier involving on-site fuel usage, on-site electricity usage, and fuel usage for transportation.

The following outputs are considered in the DEA model:

- *Delivery*: On-time delivery performance of a supplier.
- *Quality*: The ability of a supplier to meet the quality specifications determined by the company.

6.2 Numerical Example

A total of 15 suppliers are considered in the example. The values of the inputs and outputs for the example are given in Table 6.1.

When this data was used to solve the DEA example, the results presented in Table 6.2 were obtained. According to Table 6.2, suppliers S2, S5, S7, S8, S10, S11, and S13 are efficient and other suppliers are inefficient. Inefficient suppliers can improve their efficiencies by using the reference set and slack variables. For instance, supplier S1 is an inefficient supplier and its reference

TABLE 6.1

Values of Inputs and Outputs for Each Supplier

Supplier	Price (Input)	Carbon Footprint (Input)	Delivery (Output)	Quality (Output)
S1	50	255	3	1
S2	45	242	2	4
S3	50	350	1	3
S4	65	400	3	5
S5	42	245	4	2
S6	48	356	4	3
S7	53	245	5	3
S8	46	455	5	4
S9	53	345	2	5
S10	47	256	3	5
S11	54	225	4	4
S12	60	355	2	2
S13	45	235	4	2
S14	45	345	4	3
S15	55	300	1	2

TABLE 6.2

DEA Results for the Suppliers

Supplier	Efficiency (%)	Reference Set	Slacks			
			Price	Carbon	Delivery	Quality
S1	91.8	S11, S13	0	0	1	1.197
S2	100	–	0	0	0	0
S3	87	S2, S5	0	61	2	0
S4	72.3	S10	0	33.231	0	0
S5	100	–	0	0	0	0
S6	91.2	S5, S8, S10	0	35.65	0	0
S7	100	–	0	0	0	0
S8	100	–	0	0	0	0
S9	88.7	S10	0	49.943	1	0
S10	100	–	0	0	0	0
S11	100	–	0	0	0	0
S12	70	S5	0	3.5	2	0
S13	100	–	0	0	0	0
S14	97.3	S5, S8, S10	0	46.6	0	0
S15	79.7	S5, S13	0	0	3	0

set involves efficient suppliers S11 and S13. The slacks associated with two outputs, delivery and quality, are 1 and 1.197, respectively. Thus, supplier S1 should improve its delivery and quality ratings by 1 and 1.197, respectively, to become an efficient supplier. Supplier S4 is also an inefficient supplier and its reference set involves supplier S10. The slack associated with the input "carbon footprint" is 32.231. This slack value indicates that supplier S4 could be an efficient supplier if it can reduce its carbon footprint by 32.231 Mtn. CO_2.

6.3 Other Models

There are several studies that use DEA for green supplier selection. The DEA-based green supplier selection methodology proposed by Kumar et al. (2014) considered carbon footprints of suppliers as a necessary dual-role factor. Wen and Chi (2010) developed a green supplier selection methodology by integrating AHP, ANP, and DEA. Efficient candidates were determined through DEA. Then, those candidates were evaluated by using AHP and ANP. The green supplier selection methodology proposed by Kuo and Lin (2012) integrated ANP and DEA.

Evaluation and selection of third-party reverse logistic (3PL) providers using DEA is another active research area. Saen (2009) developed a DEA based on 3PL provider selection methodology that considered quantitative and qualitative data. The 3PL provider selection methodology proposed by Saen (2010) used DEA by considering multiple dual factors. Saen (2011) and Azadi and Saen (2011) considered both multiple dual factors and stochastic data in their 3PL selection methodologies. Zhou et al. (2012) evaluated third-party recyclers by developing a fuzzy confidence DEA model.

Sarkis (1999) evaluated environmentally conscious manufacturing programs by integrating ANP and DEA. Mirhedayatian et al. (2014) proposed a network DEA model for the performance evaluation of green supply chains. The various issues, including dual-role factors, undesirable outputs, and fuzzy data, were considered in the proposed model.

6.4 Conclusions

In this chapter, a model was presented for the use of DEA to solve an environmentally conscious manufacturing and product recovery (ECMPRO)-related problem. An overview of other models was also presented in the chapter.

References

Azadi M, Saen, RF A new chance-constrained data envelopment analysis for selecting third-party reverse logistics providers in the existence of dual-role factors. *Expert Systems with Applications* 2011;38: 12231–36.

Kumar A, Jain V, Kumar S. A comprehensive environment friendly approach for supplier selection. *Omega* 2014;42: 109–23.

Kuo RJ, Lin YJ. Supplier selection using analytic network process and data envelopment analysis. *International Journal of Production Research* 2012;50: 2852–63.

Mirhedayatian SM, Azadi M, et al. A novel network data envelopment analysis model for evaluating green supply chain management. *International Journal of Production Economics* 2014;147: 544–54.

Saen RF. A mathematical model for selecting third-party reverse logistics providers. *International Journal of Procurement Management* 2009;2: 180–90.

Saen RF. A new model for selecting third-party reverse logistics providers in the presence of multiple dual-role factors. *The International Journal of Advanced Manufacturing Technology* 2010;46: 405–10.

Saen RF. A decision model for selecting third-party reverse logistics providers in the presence of both dual-role factors and imprecise data. *Asia-Pacific Journal of Operational Research* 2011;28: 239–54.

Sarkis J. A methodological framework for evaluating environmentally conscious manufacturing programs. *Computers & Industrial Engineering* 1999;36: 793–810.

Wen UP, Chi JM. Developing green supplier selection procedure: A DEA approach, in: *IEEE 17th International Conference on Industrial Engineering and Engineering Management*, Xiamen, China, 2010.

Zhou H, Du G, et al. Selection of optimal third-party logistics recycler based on fuzzy DEA, in: *Proceedings of the 2012 International Conference on Automobile and Traffic Science, Materials, Metallurgy Engineering*, Wuhan, China, 2012.

7

AHP

The analytical hierarchy process (AHP) is frequently used to solve various problems of environmentally conscious manufacturing and product recovery (ECMPRO) including the evaluation of recycling, remanufacturing, reverse logistics initiatives, and ranking of suppliers based on environmental factors. In this chapter, we present an AHP model to evaluate the relative importance of various environmental criteria and to assess the relative performance of several suppliers along these criteria.

7.1 The Model

Various environmental criteria considered here are presented in the following list:

1. Waste management activities (WMA): This criterion involves the activities treating solid waste and hazardous materials.

2. Environmental management systems (EMS): This involves checking whether the supplier has a certified environmental system such as ISO 14000 certification.

3. Compliance with government regulations (CGR): The activities of suppliers must be carried out according to the environmental regulations determined by the government. This criterion may be analyzed by checking the number of fines paid by the supplier. Air and water emissions must also be checked for any possible violation.

4. Design for environment activities (DEA): The supplier must enhance product designs (ease of disassembly, modularity, etc.) to reduce the negative impact on nature.

5. Green logistics activities (GLA): The development and use of more environment-friendly logistics systems (e.g., reverse logistics) by the supplier is an important criterion.

6. Green production activities (GPA): The supplier's manufacturing process must be environmentally responsible, considering energy usage, materials, and wastes.

Comparative importance values of these criteria are presented in Table 7.1. The normalized eigenvector of the comparative importance values matrix is also presented in this table. This vector represents the relative weights given by the decision maker to the criteria.

Comparative importance values of the decision alternatives, namely, Suppliers 1, 2, 3, 4, and 5, with respect to each of the criteria, that is, WMA, EMS, CGR, DEA, GLA, and GPA, respectively, are presented in Tables 7.2 through 7.7. They also show the normalized eigenvectors of the respective comparative importance value matrices. Each of the matrices in Tables 7.2 through 7.7 has a consistency ratio (CR) whose value is less than or equal to 0.1. Table 7.8 shows the aggregate matrix of rankings of suppliers with

TABLE 7.1

Comparative Importance Values of Criteria

	WMA	**EMS**	**CGR**	**DEA**	**GLA**	**GPA**	**Norm. Eigenvector**
WMA	1	1	1/3	2	2	1/2	0.139
EMS	1	1	1/2	4	2	2	0.206
CGR	3	2	1	3	3	3	0.327
DEA	1/2	1/4	1/3	1	2	1/2	0.091
GLA	1/2	1/2	1/3	1/2	1	2	0.106
GPA	2	1/2	1/3	2	1/2	1	0.131

TABLE 7.2

Comparative Importance Values of Suppliers with Respect to WMA

	Supplier 1	**Supplier 2**	**Supplier 3**	**Supplier 4**	**Supplier 5**	**Norm. Eigenvector**
Supplier 1	1	2	1/2	5	3	0.270
Supplier 2	1/2	1	1/3	4	3	0.190
Supplier 3	2	3	1	5	2	0.365
Supplier 4	1/5	1/4	1/5	1	1/5	0.046
Supplier 5	1/3	1/3	1/2	5	1	0.129

TABLE 7.3

Comparative Importance Values of Suppliers with Respect to EMS

	Supplier 1	**Supplier 2**	**Supplier 3**	**Supplier 4**	**Supplier 5**	**Norm. Eigenvector**
Supplier 1	1	3	5	3	5	0.460
Supplier 2	1/3	1	2	1/3	1/2	0.110
Supplier 3	1/5	1/2	1	1/2	2	0.104
Supplier 4	1/3	3	2	1	3	0.225
Supplier 5	1/5	2	1/2	1/3	1	0.100

TABLE 7.4

Comparative Importance Values of Suppliers with Respect to CGR

	Supplier 1	Supplier 2	Supplier 3	Supplier 4	Supplier 5	Norm. Eigenvector
Supplier 1	1	1/5	2	3	2	0.194
Supplier 2	5	1	4	3	3	0.482
Supplier 3	1/2	1/4	1	2	2	0.137
Supplier 4	1/3	1/3	1/2	1	1	0.091
Supplier 5	1/2	1/3	1/2	1	1	0.097

TABLE 7.5

Comparative Importance Values of Suppliers with Respect to DEA

	Supplier 1	Supplier 2	Supplier 3	Supplier 4	Supplier 5	Norm. Eigenvector
Supplier 1	1	2	5	1/2	3	0.305
Supplier 2	1/2	1	2	2	3	0.250
Supplier 3	1/5	1/2	1	1/2	2	0.103
Supplier 4	2	1/2	2	1	4	0.273
Supplier 5	1/3	1/3	1/2	1/4	1	0.069

TABLE 7.6

Comparative Importance Values of Suppliers with Respect to GLA

	Supplier 1	Supplier 2	Supplier 3	Supplier 4	Supplier 5	Norm. Eigenvector
Supplier 1	1	3	1/3	4	3	0.274
Supplier 2	1/3	1	1/2	3	4	0.188
Supplier 3	3	2	1	4	3	0.388
Supplier 4	1/4	1/3	1/4	1	1/2	0.062
Supplier 5	1/3	1/4	1/3	2	1	0.089

TABLE 7.7

Comparative Importance Values of Suppliers with Respect to GPA

	Supplier 1	Supplier 2	Supplier 3	Supplier 4	Supplier 5	Norm. Eigenvector
Supplier 1	1	3	1/2	5	4	0.321
Supplier 2	1/3	1	1/3	2	4	0.161
Supplier 3	2	3	1	3	2	0.349
Supplier 4	1/5	1/2	1/3	1	2	0.093
Supplier 5	1/4	1/4	1/2	1/2	1	0.077

TABLE 7.8

Aggregate of Ratings of Suppliers

	WMA	EMS	CGR	DEA	GLA	GPA
Supplier 1	0.270	0.460	0.194	0.305	0.274	0.321
Supplier 2	0.190	0.110	0.482	0.250	0.188	0.161
Supplier 3	0.365	0.104	0.137	0.103	0.388	0.349
Supplier 4	0.046	0.225	0.091	0.273	0.062	0.093
Supplier 5	0.129	0.100	0.097	0.069	0.089	0.077

respect to each criterion. This matrix is the aggregate of the eigenvectors obtained in Tables 7.2 through 7.7.

The following normalized ranks for the suppliers are obtained by multiplying the aggregate matrix in Table 7.8 with the normalized eigenvector presented in Table 7.1: Supplier 1: 0.296, Supplier 2: 0.259, Supplier 3: 0.223, Supplier 4: 0.128, Supplier 5: 0.095. It is clear that Supplier 1 has the highest ranking.

7.2 Other Models

There have been many applications of AHP in the design for the environment. The environmental performance of alternative product designs were evaluated in Azzone and Noci (1996) using AHP. Choi et al. (2008) presented a comparative analysis of five designs for environment strategies based on the results of an AHP analysis. The AHP-based green product design methodology developed by Wang et al. (2012) prevented designers from conducting a detailed analysis (e.g., life cycle assessment) for every product design alternative. The recycling potential of materials is evaluated by Kim et al. (2009) through an AHP analysis based on environmental and economic factors.

The use of AHP in sustainable evaluation of suppliers is another active research area. Noci (1997) employed AHP to develop a green vendor rating system. The AHP-based supplier evaluation methodology proposed by Handfield et al. (2002) assessed several suppliers considering environmental issues. The sustainable supplier assessment methodology proposed by Dai and Blackhurst (2012) involved four phases. First, quality function deployment was employed to link customer requirements with the company's sustainability strategy. Then, the sustainable purchasing competitive priority was determined. Next, the criteria required for the sustainability assessment of suppliers were developed. Finally, suppliers were evaluated using AHP.

Barker and Zabinsky (2011) obtained a preference ordering of eight alternative reverse logistics network configurations by using AHP with sensitivity analysis. Jiang et al. (2012) employed AHP to solve the remanufacturing portfolio selection problem. The sustainability level of manufacturing systems was evaluated in Ziout et al. (2013) based on the results of an AHP analysis. Sarmiento and Thomas (2010) used AHP to determine the improvement areas in the implementation of green initiatives. Shaik and Abdul-Kader (2012) developed a two-phase performance measurement system for reverse logistic systems. The first phase involved the development of a reverse logistics performance measurement system by integrating balanced scorecard and performance prism. The overall comprehensive performance index (OCPI) was calculated in the second phase based on the integration of AHP with the system constructed in the first phase. The Reman decision making framework (RDMF) developed in Subramoniam et al. (2010) was validated in Subramoniam et al. (2013) by using AHP.

7.3 Conclusions

In this chapter, a model was presented for the use of AHP to solve an ECMPRO-related problem. An overview of other models was also presented in the chapter.

References

Azzone G, Noci G. Measuring the environmental performance of new products: An integrated approach. *International Journal of Production Research* 1996;34: 3055–78.

Barker TJ, Zabinsky ZB. A multicriteria decision making model for reverse logistics using analytical hierarchy process. *Omega* 2011;39: 558–73.

Choi JK, Nies LF, Ramani K. A framework for the integration of environmental and business aspects toward sustainable product development. *Journal of Engineering Design* 2008;19: 431–46.

Dai J, Blackhurst J. A four-phase AHP-QFD approach for supplier assessment: A sustainability perspective. *International Journal of Production Research* 2012;50: 5474–90.

Handfield R, Walton SV, Sroufe R, Melnyk SA. Applying environmental criteria to supplier assessment: A study in the application of the analytical hierarchy process. *European Journal of Operational Research* 2002;141: 70–87.

Jiang Z, Zhang H, Sutherland JW. Development of multi-criteria decision making model for remanufacturing technology portfolio selection. *Journal of Cleaner Production* 2012;19: 1939–45.

Kim J, Hwang Y, Park K. An assessment of the recycling potential of materials based on environmental and economic factors; case study in South Korea. *Journal of Cleaner Production* 2009;17: 1264–71.

Noci G. Designing a "green" vendor rating systems for the assessment of a supplier's environmental performance. *European Journal of Purchasing & Supply Management* 1997;3: 103–14.

Sarmiento R, Thomas A. Identifying improvement areas when implementing green initiatives using a multitier AHP approach. *Benchmarking: An International Journal* 2010;17: 452–63.

Shaik M, Abdul-Kader W. Performance measurement of reverse logistics enterprise: a comprehensive and integrated approach. *Measuring Business Excellence* 2012;16: 23–34.

Subramoniam R, Huisingh D, Chinnam RB. Aftermarket remanufacturing strategic planning decision-making framework: Theory and practice. *Journal of Cleaner Production* 2010;18: 1575–86.

Subramoniam R, Huisingh D, Chinnam RB, Subramoniam S. Remanufacturing decision-making framework (RDMF): Research validation using the analytical hierarchical process. *Journal of Cleaner Production* 2013;40: 212–20.

Wang X, Chan HK, Li D. A case study of AHP based model for green product design selection, in: *Proceedings of the EWG-DSS Liverpool-2012 Workshop on Decision Support Systems and Operations Management Trends and Solutions in Industries*, Liverpool, 2012, pp. 1–6.

Ziout A, Azab A, Altarazi S, ElMaraghy WH. Multi-criteria decision support for sustainability assessment of manufacturing system reuse. *CIRP Journal of Manufacturing Science and Technology* 2013;6: 59–69.

8

Fuzzy AHP

Although the analytical hierarchy process (AHP) is a sophisticated multiple criteria decision making (MCDM) tool with a well-defined comparison scale, it is often criticized for its inability to consider the vagueness and uncertainty of judgments of the decision maker. Fuzzy logic is often incorporated into the pairwise comparison process to address the abovementioned deficiency of AHP, referred to as fuzzy AHP (FAHP). FAHP is one of the most widely used MCDM techniques to deal with the high level of uncertainty associated with environmentally conscious manufacturing and product recovery (ECMPRO)-related problems. In this chapter, we present a FAHP-based model that evaluates and ranks in order alternative product recovery facilities (PRFs).

8.1 The Model

The hierarchy of the FAHP model to evaluate the potential PRFs is presented in Figure 8.1. As is clear from the figure, the first level involves the goal: selection of the potential facilities from a set of candidate recovery facilities. The second level contains the evaluation criteria of the candidate recovery facilities involved in the last level. The fixed cost (FC) of a facility is a commonly used criterion in forward supply chain. The remaining four criteria, which are peculiar to product recovery, are explained in the following list:

Customer service (CS): Customer service associated with a PRF can be evaluated based on several factors. First of all, the facility must comply with the environmental regulations. Second, government incentives must be utilized effectively. Third, the type and quantity of the incentives given to collection centers that supply the used products and customers who buy reprocessed products must be determined effectively.

Multiplication of average disassembly time (DT) by throughput (TP): DT directly affects the TP of a PRF. A high DT may result in a low TP, while a high TP may be experienced for a low DT. The multiplication of DT by TP is used as a criterion to compensate for high or low supply of used products.

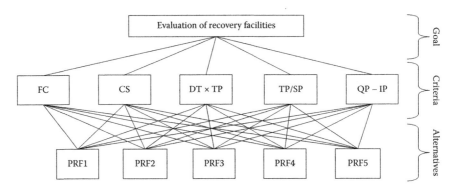

FIGURE 8.1
Hierarchical structure for fuzzy AHP.

> *Ratio of throughput (TP) to supply of used products (SP)*: SP is another factor directly affecting the TP of a PRF. A high SP may cause a high TP, while a low SP may lead to a low TP. The ratio of TP to SP is used as a criterion to compensate for high or low SP.

> *Difference between the average quality of reprocessed products (QP) and the average quality of incoming used products (IP)*: The difference between QP and IP is used as a criterion due to high-level correlation between QP and IP.

8.2 Numerical Example

Linguistic rating is used to represent comparative importance values, which are converted into triangular fuzzy numbers (TFNs). TFNs in Table 8.1 are employed to quantify different linguistic ratings.

Comparative importance values of the various criteria are presented in Table 8.2. The normalized eigenvector of the comparative importance values matrix is also presented in this table. This vector represents the relative weights given by the decision maker to the criteria.

Comparative importance values of the decision alternatives, that is, suppliers PRF1, PRF2, PRF3, PRF4, and PRF5, with respect to each of the criteria, namely, FC, CS, DT × TP, TP/SP, and QP – IP, respectively, are presented in Tables 8.3 through 8.7. They also show the normalized eigenvectors of the respective comparative importance value matrices. Each of the matrices in Tables 8.3 through 8.7 has a consistency ratio (CR) whose value is less than or equal to 0.1. Table 8.8 shows the aggregate matrix of rankings of suppliers with respect to each criterion. This matrix is the aggregate of the eigenvectors obtained in Tables 8.3 through 8.7.

TABLE 8.1

Conversion of Linguistic Preferences into Triangular Fuzzy Numbers

Linguistic Scale	Fuzzy Triangular Scale	Reciprocal Fuzzy Triangular Scale
Equally preferred (E)	(1, 1, 1)	(1, 1, 1)
Moderately preferred (M)	(2, 3, 4)	(1/4, 1/3, 1/2)
Strongly preferred (S)	(4, 5, 6)	(1/6, 1/5, 1/4)
Very strongly preferred (VS)	(6, 7, 8)	(1/8, 1/7, 1/6)
Absolutely preferred (A)	(9, 9, 9)	(1/9, 1/9, 1/9)

TABLE 8.2

Comparative Importance Values of Criteria

	FC	CS	DT × TP	TP/SP	QP − IP	Normalized Eigenvector
FC	E	S	S	M	S	0.482291
CS	1/S	E	1/M	E	M	0.116662
DT × TP	1/S	M	E	E	S	0.195711
TP/SP	1/M	1/E	1/E	E	M	0.148647
QP − IP	1/S	1/M	1/S	1/M	E	0.056689

TABLE 8.3

Comparative Importance Values of Recovery Facilities with Respect to FC

	FC					
	PRF1	PRF2	PRF3	PRF4	PRF5	Normalized Eigenvector
PRF1	E	M	S	S	M	0.382901
PRF2	1/M	E	VS	A	M	0.302226
PRF3	1/S	1/VS	E	VS	S	0.170368
PRF4	1/S	1/A	1/VS	E	M	0.075578
PRF5	1/M	1/M	1/S	1/M	E	0.068926

TABLE 8.4

Comparative Importance Values of Recovery Facilities with Respect to CS

	CS					
	PRF1	PRF2	PRF3	PRF4	PRF5	Normalized Eigenvector
PRF1	E	S	VS	M	S	0.437994
PRF2	1/S	E	1/VS	M	S	0.136188
PRF3	1/VS	VS	E	S	VS	0.289787
PRF4	1/M	1/M	1/S	E	M	0.092300
PRF5	1/S	1/S	1/VS	1/M	E	0.043731

TABLE 8.5

Comparative Importance Values of Recovery Facilities with
Respect to DT × TP

			DT × TP			
	PRF1	**PRF2**	**PRF3**	**PRF4**	**PRF5**	**Normalized Eigenvector**
PRF1	E	M	S	VS	VS	0.472282
PRF2	1/M	E	S	M	S	0.252203
PRF3	1/S	1/S	E	S	S	0.157503
PRF4	1/VS	1/M	1/S	E	M	0.075580
PRF5	1/VS	1/S	1/S	1/M	E	0.042432

TABLE 8.6

Comparative Importance Values of Recovery Facilities with
Respect to TP/SP

			TP/SP			
	PRF1	**PRF2**	**PRF3**	**PRF4**	**PRF5**	**Normalized Eigenvector**
PRF1	E	VS	1/S	1/S	1/M	0.189083
PRF2	1/VS	E	M	VS	S	0.320935
PRF3	S	1/M	E	1/M	S	0.168606
PRF4	S	1/VS	M	E	VS	0.254163
PRF5	M	1/S	1/S	1/VS	E	0.067213

TABLE 8.7

Comparative Importance Values of Recovery Facilities with
Respect to QP – IP

			QP – IP			
	PRF1	**PRF2**	**PRF3**	**PRF4**	**PRF5**	**Normalized Eigenvector**
PRF1	E	E	S	1/M	1/M	0.136266
PRF2	1/E	E	M	VS	1/VS	0.196012
PRF3	1/S	1/M	E	S	1/S	0.108546
PRF4	M	1/VS	1/S	E	M	0.220888
PRF5	M	VS	S	1/M	E	0.338289

TABLE 8.8

Aggregate of Ratings of Product Recovery Facilities

	FC	CS	DT × TP	TP/SP	QP – IP
PRF1	0.382901	0.437994	0.472282	0.189083	0.136266
PRF2	0.302226	0.136188	0.252203	0.320935	0.196012
PRF3	0.170368	0.289787	0.157503	0.168606	0.108546
PRF4	0.075578	0.092300	0.075580	0.254163	0.220888
PRF5	0.068926	0.043731	0.042432	0.067213	0.338289

Multiplying the matrix in Table 8.8 with the normalized eigenvector obtained in Table 8.2, the following normalized ranks for the facilities are obtained: $Rank_{PRF1}$: 0.3640, $Rank_{PRF2}$: 0.2698, $Rank_{PRF3}$: 0.1780, $Rank_{PRF4}$: 0.1123, $Rank_{PRF5}$: 0.0758. It is clear that PRF1 has the highest ranking.

8.3 Other Models

Kuo et al. (2006) developed an environmentally conscious design methodology by integrating AHP and fuzzy multiattribute decision making. In Li et al. (2008), AHP and fuzzy logic were integrated to develop an optimal modular formulation for modular product design considering environmental issues. Yu et al. (2000) considered three environmental criteria (i.e., environmental impact, recycling associated cost, and recoverable material content) in their FAHP-based methodology developed for the determination of the most appropriate recycling option for end-of-life (EOL) products.

There are many studies that used FAHP for the environmental evaluation of suppliers: Lu et al. (2007), Lee et al. (2009), Grisi et al. (2010), Çiftçi and Büyüközkan (2011), and Amin and Zhang (2012). Lee et al. (2012) analyzed the criteria used for green supplier selection in the Taiwanese hand tool industry by developing an FAHP-based methodology. Chiou et al. (2008) compared the green supply-chain management (GSCM) practices of American, Japanese, and Taiwanese electronics manufacturers operating in China by using FAHP. Chiou et al. (2012) employed FAHP to determine the most important criteria in reverse logistics implementation. Efendigil et al. (2008) developed a methodology of third-party reverse logistics provider selection by integrating FAHP and artificial neural networks.

Potential facilities in a set of candidate recovery facilities operating in the region were determined in Gupta and Nukala (2005) using FAHP. The effective factors associated with the sustainable supply-chain management in publishing industry were analyzed in Shaverdi et al. (2013) by developing an FAHP-based approach.

8.4 Conclusions

In this chapter, a model was presented for the use of FAHP to solve an ECMPRO-related problem. The model involved the evaluation of alternative PRFs. An overview of other models was also presented in the chapter.

References

Amin SH, Zhang G. An integrated model for closed-loop supply chain configuration and supplier selection: Multi-objective approach. *Expert Systems with Applications* 2012; 39: 6782–91.

Chiou CY, Chen HC, Yu CT, Yeh CY. Consideration factors of reverse logistics implementation: A case study of Taiwan's electronics industry. *Procedia: Social and Behavioral Sciences* 2012;40: 375–81.

Chiou CY, Hsu CW, Hwang WY. Comparative investigation on green supplier selection of the American, Japanese and Taiwanese Electronics Industry in China, in: *IEEE International Conference on Industrial Engineering and Engineering Management*, Singapore, 2008, pp. 1909–14.

Çiftçi G, Büyüközkan G. A fuzzy MCDM approach to evaluate green suppliers. *International Journal of Computational Intelligence Systems* 2011;4: 894–909.

Efendigil T, Onut S, Kongar E. A holistic approach for selecting a third-party reverse logistics provider in the presence of vagueness. *Computers & Industrial Engineering* 2008;54: 269–87.

Grisi R, Guerra L, Naviglio G. Supplier performance evaluation for green supply chain management, in: *Business Performance Measurement and Management*. Springer Berlin Heidelberg, 2010, pp. 149–63.

Gupta SM, Nukala S. A fuzzy AHP-based approach for selecting potential recovery facilities in a closed loop supply chain, in: *Proceedings of the SPIE International Conference on Environmentally Conscious Manufacturing V*, SPIE, Boston, MA, 2005, pp. 58–63.

Kuo T-C, Chang S-H, Huang S. Environmentally conscious design by using fuzzy multi-attribute decision-making. *The International Journal of Advanced Manufacturing Technology* 2006;29: 419–25.

Lee AHI, Kang H-Y, Hsu C-F, Hung H-C. A green supplier selection model for high-tech industry. *Expert Systems with Applications* 2009;36: 7917–27.

Lee T-R, Le TPN, Genovese A, Koh LS. Using FAHP to determine the criteria for partner's selection within a green supply chain: The case of hand tool industry in Taiwan. *Journal of Manufacturing Technology Management* 2012;23: 25–55.

Li J, Zhang H-C, Gonzalez MA, Yu S. A multi-objective fuzzy graph approach for modular formulation considering end-of-life issues. *International Journal of Production Research* 2008;46: 4011–33.

Lu LYY, Wu CH, Kuo TC. Environmental principles applicable to green supplier evaluation by using multi-objective decision analysis. *International Journal of Production Research* 2007;45: 4317–31.

Shaverdi M, Heshmati MR, Eskandaripour E, Tabar AAA. Developing sustainable SCM evaluation model using fuzzy AHP in publishing industry. *Procedia Computer Science* 2013;17: 340–9.

Yu Y, Jin K, Zhang HC, Ling FF, Barnes D. A decision-making model for materials management of end-of-life electronic products. *Journal of Manufacturing Systems* 2000;19: 94–107.

9

Analytic Network Process

The analytic network process (ANP) models decision problems using a network structure that allows for the consideration of dependencies and feedback among decision criteria. Due to its ability in modeling complex systems, ANP has been used to solve various environmentally conscious manufacturing and product recovery (ECMPRO)-related issues including the evaluation of ECMPRO programs, evaluation and selection of third-party reverse logistics providers, evaluation of return centers, and evaluation of connection types in design for disassembly. In this chapter, we present a model where ANP is used to calculate the performance indices (efficiency scores) of candidate collection centers, with respect to qualitative criteria taken from the perspective of a remanufacturing facility interested in buying used products from the collection centers.

9.1 The Model

The problem of evaluating the efficiencies of the candidate collection centers is framed as a four-level hierarchy, as shown in Figure 9.1. The first level contains the objective of evaluation of the candidate collection centers, the second level consists of the main evaluation criteria taken from the perspective of a remanufacturing facility, the third level contains the subcriteria under each main criterion, and the fourth level contains the candidate collection centers. The main and subcriteria considered are as follows:

- *Reliability*: This criterion relates to a collection center delivering the used products at the right time, at the right remanufacturing facility, in the right quantity, and in the promised condition. The subcriteria considered under this main criterion are: (a) delivery reliability and (b) conformance to standards.

- *Capability/responsiveness*: This criterion reflects the velocity at which a collection center supplies the used products to the remanufacturing facility and the collection center's ability to adapt to sudden demand fluctuations. The subcriteria considered here are: (a) order fulfillment lead time, (b) flexibility in adapting to demand fluctuations, and (c) design capabilities.

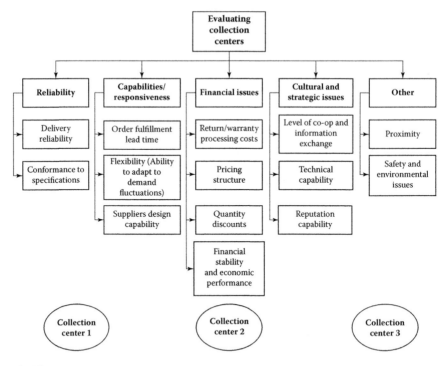

FIGURE 9.1
Hierarchical structure for ANP.

- *Financial issues*: This criterion reflects the costs and other financial aspects involved. The subcriteria are: (a) returns/warranty processing costs, (b) pricing structure, (c) quantity discounts, and (d) financial stability and economic performance of the collection center.
- *Cultural and strategic issues*: This criterion consists of the following subcriteria: (a) level of cooperation and information exchange between the collection center and the remanufacturing facility, (b) collection center's reputation in the industry, and (c) collection center's technical capability (how knowledgeable the collection center is about the product).
- *Others*: This criterion considers miscellaneous aspects that are not considered in the other criteria. They are: (a) proximity of the collection center to the remanufacturing facility and (b) safety and environmental issues.

Three collection centers, S1, S2, and S3, are considered in this example. Table 9.1 shows the pairwise comparison matrix for the main criteria and also the normalized eigenvector of the matrix. The elements of the normalized eigenvector are the impacts given to the main criteria with respect to the objective.

TABLE 9.1

Comparative Importance Values of Main Criteria

Criteria	Reliability	Responsiveness	Financial Issues	Cultural and Strategic Issues	Others	Normalized Eigenvector
Reliability	1	1/4	3	1	1/6	0.113
Responsiveness	4	1	5	3	3	0.426
Financial Issues	1/3	1/5	1	1/2	1/3	0.064
Cultural and Strategic Issues	1	1/3	2	1	1/3	0.114
Others	6	1/3	3	3	1	0.284

Tables 9.2 through 9.6 show the pairwise comparison matrices of subcriteria with respect to their main criteria, and also the corresponding normalized eigenvectors of the matrices. Note that each of the matrices in Tables 9.1 through 9.6 has a consistency ratio (CR) of less than or equal to 0.1.

Table 9.7 shows the matrix of interdependencies (called the supermatrix M) among the subcriteria with respect to their main criteria. This supermatrix M is made to converge to obtain a long-term stable set of impacts. For convergence, M must be made column stochastic, which is done by raising M to the power of 2^{k+1}, where k is an arbitrarily large number. In this example, $k = 59$. Table 9.8 shows the converged supermatrix.

Table 9.9 shows the relative ratings of the candidate collection centers, S1, S2, and S3, with respect to the subcriteria. These ratings are obtained after carrying out pairwise comparisons between the candidate collection centers with respect to the subcriteria and then obtaining the normalized eigenvector.

To obtain pairwise comparisons, interdependencies, and relative ratings (see Tables 9.2 through 9.9), decision makers could be invited to participate in relevant survey questionnaires, individual interviews, focus groups, and on-site observations.

Table 9.10 shows the desirability index calculated for each candidate collection center.

The overall performance index for each of the three collection centers is calculated by multiplying the desirability index (Table 9.10) of each collection center for each main criterion by the impact of that criterion (Table 9.1),

TABLE 9.2

Comparative Importance Values of Subcriteria under "Reliability"

Subcriteria	Delivery Reliability (DR)	Conformance to Specs (CS)	Norm. Eigenvector
Delivery Reliability (DR)	1	1/3	0.25
Conformance to Specs (CS)	3	1	0.75

TABLE 9.3

Comparative Importance Values of Subcriteria under "Responsiveness"

Subcriteria	Order Fulfillment Lead Time (OFT)	Flexibility (F)	Design Capability (DC)	Norm. Eigenvector
Order Fulfillment Lead Time (OFT)	1	3	4	0.623
Flexibility (F)	1/3	1	2	0.239
Design Capability (DC)	1/4	1/2	1	0.137

TABLE 9.4

Comparative Importance Values of Subcriteria under "Financial Issues"

Subcriteria	Returns/Warranty Processing Costs (RC)	Pricing Structure (PS)	Qty Discounts (QD)	Economic Performance and Financial Stability (EP)	Norm. Eigenvector
Returns/Warranty Processing Costs (RC)	1	1/3	1/2	1/4	0.099
Pricing Structure (PS)	3	1	2	1	0.345
Qty Discounts (QD)	2	1/2	1	1/2	0.185
Economic Performance and Financial Stability (EP)	4	1	2	1	0.37

TABLE 9.5

Comparative Importance Values of Subcriteria under "Cultural and Strategic Issues"

Subcriteria	Level of Co-Op and Info. Exchange (*Co-op*)	Technical Capability (*TC*)	Reputation (*R*)	Norm. Eigenvector
Level of Co-Op and Info. Exchange (*Co-op*)	1	5	6	0.722
Technical Capability (*TC*)	1/5	1	2	0.174
Reputation (*R*)	1/6	1/2	1	0.103

TABLE 9.6

Comparative Importance Values of Subcriteria under "Others"

Subcriteria	Proximity (*P*)	Safety and Environment (*SE*)	Norm. Eigenvector
Proximity (*P*)	1	1/3	0.25
Safety and Environment (*SE*)	3	1	0.75

and summing over all the criteria. Table 9.11 shows the overall performance indices (efficiencies) for the three collection centers. It is clear that S3 has the highest ranking.

9.2 Other Models

Researchers have developed several ANP-based methodologies for various ECMPRO-related issues. Sarkis (1999) integrated ANP and data envelopment analysis for the evaluation of environmentally conscious manufacturing programs. Meade and Sarkis (2002) developed an ANP-based methodology for the evaluation and selection of third-party reverse logistics providers. Ravi et al. (2005) integrated ANP and balanced score card to determine the best reverse logistics option for end-of-life (EOL) computers. Gungor (2006) evaluated different connection types in design for disassembly using ANP. Hsu and Hu (2009) integrated hazardous substance management into the supplier selection process by developing an ANP-based supplier selection methodology. Hsu et al. (2011) extended it by constructing a sustainability balanced scorecard. Chen et al. (2012) used ANP to solve the strategy selection problem of a green supply chain.

TABLE 9.7

Matrix of Interdependencies (Supermatrix M)

	DR	CS	OFT	F	DC	RC	PS	QD	EP	Co-op	TC	R	P	SE
DR	0	1	0	0	0	0	0	0	0	0	0	0	0	0
CS	1	0	0	0	0	0	0	0	0	0	0	0	0	0
OFT	0	0	0	0.8	0.75	0	0	0	0	0	0	0	0	0
F	0	0	0.75	0	0.25	0	0	0	0	0	0	0	0	0
DC	0	0	0.25	0.2	0	0	0	0	0	0	0	0	0	0
RC	0	0	0	0	0	0	0.29	0.13	0.201	0	0	0	0	0
PS	0	0	0	0	0	0.53	0	0.62	0.6	0	0	0	0	0
QD	0	0	0	0	0	0.29	0.16	0	0.11	0	0	0	0	0
EP	0	0	0	0	0	0.163	0.53	0.23	0	0	0	0	0	0
Co-op	0	0	0	0	0	0	0	0	0	0	0.75	0.83	0	0
TC	0	0	0	0	0	0	0	0	0	0.8	0	0.16	0	0
R	0	0	0	0	0	0	0	0	0	0.2	0.25	0	0	0
P	0	0	0	0	0	0	0	0	0	0	0	0	0	1
SE	0	0	0	0	0	0	0	0	0	0	0	0	1	0

TABLE 9.8

Converged Supermatrix

Subcriteria	Stabilized Relative Impact
Delivery reliability	1
Conformance to specs	1
Order fulfillment lead time	0.44
Flexibility	0.38
Design capability	0.18
Returns/warranty	0.19
Pricing structure	0.38
Qty discounts	0.15
Economic performance and financial stability	0.27
Co-op and info. exchange	0.44
Technical capability	0.38
Reputation	0.18
Proximity	1
Safety and environment	1

TABLE 9.9

Relative Ratings of Collection Centers with respect to Subcriteria

Subcriteria/Alternate Collection Centers	S1	S2	S3
DR	0.33	0.141	0.524
CS	0.345	0.543	0.11
OFT	0.274	0.068	0.657
F	0.109	0.309	0.581
DC	0.09	0.25	0.652
RC	0.681	0.216	0.102
PS	0.309	0.581	0.109
QD	0.376	0.151	0.471
EP	0.137	0.623	0.239
Co-op	0.33	0.075	0.59
TC	0.137	0.239	0.623
R	0.67	0.23	0.12
P	0.292	0.092	0.615
SE	0.137	0.239	0.623

ANP was used in this scorecard to determine the relative weights of the selected measures. Vinodh et al. (2012) evaluated the sustainable business practices of an Indian company by using ANP. Govindan et al. (2013) proposed a two-stage third-party reverse logistic provider selection methodology. AHP was used in the first stage to carry out an initial screening

TABLE 9.10

Desirability Indices

Criteria/Collection Centers	S1	S2	S3
Reliability	0.3429	0.4432	0.2138
Responsiveness	0.0874	0.0528	0.2488
Financial issues	0.0783	0.148551	0.054
Cultural and strategic issues	0.1271	0.0438	0.2299
Others	0.176	0.2027	0.6211

TABLE 9.11

Overall Performance Indices

Collection Center	Performance Index
S1	0.2318
S2	0.2322
S3	0.5359

of criteria. The selection of third-party reverse logistics providers was achieved in the second stage using ANP. Dou et al. (2014) integrated ANP and gray relational analysis to determine effective green supplier development programs.

High-level ambiguity and vagueness associated with ECMPRO systems forced researchers to integrate ANP with fuzzy logic. Tuzkaya et al. (2009) evaluated the environmental performance of suppliers by integrating fuzzy ANP and fuzzy PROMETHEE. Büyüközkan and Çiftçi (2011) developed a fuzzy ANP-based methodology to evaluate the sustainability level of suppliers. Büyüközkan and Çiftçi (2012) evaluated the green supply-chain management practices of an automotive factory located in Turkey using fuzzy ANP. Bhattacharya et al. (2014) integrated fuzzy ANP and balanced score card to develop a green supply-chain performance measurement system.

9.3 Conclusions

In this chapter, a model was presented for the use of ANP to solve an ECMPRO-related problem. An overview of other models was also presented in the chapter.

References

Bhattacharya A, Mohapatra P, Kumar V, Dey PK, Brady M, Tiwari MK, Nudurupati SS. Green supply chain performance measurement using fuzzy ANP-based balanced scorecard: a collaborative decision-making approach. *Production Planning & Control* 2014;25: 698–714.

Büyüközkan G, Çiftçi G. A novel fuzzy multi-criteria decision framework for sustainable supplier selection with incomplete information. *Computers in Industry* 2011;62: 164–74.

Büyüközkan G, Çiftçi G. Evaluation of the green supply chain management practices: A fuzzy ANP approach. *Production Planning & Control* 2012;23: 405–18.

Chen C-C, Shih H-S, Shyur H-J, Wu K-S. A business strategy selection of green supply chain management via an analytic network process. *Computers & Mathematics with Applications* 2012;64: 2544–57.

Dou Y, Zhu Q, Sarkis J. Evaluating green supplier development programs with a grey-analytical network process-based methodology. *European Journal of Operational Research* 2014;233: 420–31.

Govindan K, Sarkis J, Palaniappan M. An analytic network process-based multicriteria decision making model for a reverse supply chain. *The International Journal of Advanced Manufacturing Technology* 2013;68: 863–80.

Gungor A. Evaluation of connection types in design for disassembly (DFD) using analytic network process. *Computers & Industrial Engineering* 2006;50: 35–54.

Hsu C-W, Hu AH. Applying hazardous substance management to supplier selection using analytic network process. *Journal of Cleaner Production* 2009;17: 255–64.

Hsu C-W, Hu AH, Chiou C-Y, Chen T-C. Using the FDM and ANP to construct a sustainability balanced scorecard for the semiconductor industry. *Expert Systems with Applications* 2011;38: 12891–9.

Meade L, Sarkis J. A conceptual model for selecting and evaluating third-party reverse logistics providers. *Supply Chain Management: An International Journal* 2002;7: 283–95.

Ravi V, Shankar R, Tiwari MK. Analyzing alternatives in reverse logistics for end-of-life computers: ANP and balanced scorecard approach. *Computers & Industrial Engineering* 2005;48: 327–56.

Sarkis J. A methodological framework for evaluating environmentally conscious manufacturing programs. *Computers & Industrial Engineering* 1999;36: 793–810.

Tuzkaya G, Ozgen A, Ozgen D, Tuzkaya UR. Environmental performance evaluation of suppliers: A hybrid fuzzy multi criteria decision approach. *International Journal of Environment Science and Technology* 2009;6: 477–90.

Vinodh S, Prasanna M, Manoj S. Application of analytical network process for the evaluation of sustainable business practices in an Indian relays manufacturing organization. *Clean Technologies and Environmental Policy* 2012;14: 309–17.

10

DEMATEL

The decision making trial and evaluation laboratory (DEMATEL) method is employed to determine causal relationships among evaluation criteria. It also helps researchers understand the influence level of each criterion. The use of DEMATEL in environmentally conscious manufacturing and product recovery (ECMPRO)-related problems is limited to only a few studies. Evaluation of green supply-chain management practices and vendor selection for the recycled material are some of the issues considered in these studies. In this chapter, we present a DEMATEL-based model. The model uses DEMATEL to evaluate various green supply-chain management practices.

10.1 The Model

In this section, we present a DEMATEL model to evaluate the relationships among green supply-chain practices, organizational performance, and external driving factors.

10.1.1 Determination of Criteria

10.1.1.1 Green Supply-Chain Practices

Green purchasing (CR1): Green purchasing involves the formal introduction and integration of environmental issues and concerns into the purchasing process.

Green design (CR2): Green design involves the consideration of certain environmental criteria in the product design process.

Product recovery (CR3): Product recovery involves the minimization of the amount of waste sent to landfills by recovering materials and parts from returned or end-of-life (EOL) products via recycling and remanufacturing.

Green information systems (CR4): Green information systems involve the use of information systems to achieve environmental objectives.

10.1.1.2 Organizational Performance

Environmental performance (CR5): Environmental performance relates the ability of manufacturing plants to reduce air emissions, effluent waste, and solid wastes, and the ability to decrease consumption of hazardous and toxic materials.

Economic performance (CR6): Economic performance relates to the manufacturing plant's ability to reduce costs associated with purchased materials, energy consumption, waste treatment, waste discharge, and fines for environmental accidents.

Operational performance (CR7): Operational performance relates to the manufacturing plant's capabilities to more efficiently produce and deliver products to customers.

10.1.1.3 External Driving Factors

Regulations (CR8): There are three types of environmental regulations: domestic, government, and international.

Pressures from stakeholders (CR9): Stakeholders (shareholders, customers, etc.) force companies to implement green supply-chain initiatives.

10.1.2 Application of DEMATEL Methodology

The following steps show the example data and implementation of DEMATEL methodology:

Step 1: *Construct the direct relation matrix*: Using the fuzzy linguistic scale defined in Table 10.1, the direct relation matrix is formed as shown in Table 10.2.

Step 2: *Normalize the direct relation matrix*: Obtain the normalized direct relation matrix as presented in Table 10.3.

Step 3: *Obtain the total relation matrix*: Obtain the total relation matrix as shown in Table 10.4.

Step 4: *Construct a cause and effect diagram*: The cause and effect diagram is given in Figure 10.1.

TABLE 10.1

Fuzzy Linguistic Scale

Linguistic Variable	Influence Score	Corresponding Triangular Fuzzy Numbers (TFNs)
No influence	0	(0, 0.1, 0.3)
Very low influence	1	(0.1, 0.3, 0.5)
Low influence	2	(0.3, 0.5, 0.7)
High influence	3	(0.5, 0.7, 0.9)
Very high influence	4	(0.7, 0.9, 1)

TABLE 10.2

Direct Relation Matrix

	CR1	CR2	CR3	CR4	CR5	CR6	CR7	CR8	CR9
CR1	0	2	1	1	4	3	2	0	2
CR2	2	0	3	1	4	2	2	0	2
CR3	1	1	0	1	4	2	1	0	2
CR4	2	0	2	0	4	1	1	0	2
CR5	2	2	2	2	0	3	3	0	4
CR6	1	1	1	1	2	0	3	0	2
CR7	0	0	0	0	2	2	0	0	2
CR8	3	3	4	3	4	2	2	0	3
CR9	3	3	4	3	4	3	3	0	0

TABLE 10.3

Normalized Direct Relation Matrix

	CR1	CR2	CR3	CR4	CR5	CR6	CR7	CR8	CR9
CR1	0	0.07	0.04	0.04	0.14	0.11	0.07	0	0.07
CR2	0.07	0	0.11	0.04	0.14	0.07	0.07	0	0.07
CR3	0.04	0.04	0	0.04	0.14	0.07	0.04	0	0.07
CR4	0.07	0	0.07	0	0.14	0.04	0.04	0	0.07
CR5	0.07	0.07	0.07	0.07	0	0.11	0.11	0	0.14
CR6	0.04	0.04	0.04	0.04	0.07	0	0.11	0	0.07
CR7	0	0	0	0	0.07	0.07	0	0	0.07
CR8	0.11	0.11	0.14	0.11	0.14	0.07	0.07	0	0.11
CR9	0.11	0.11	0.14	0.11	0.14	0.11	0.11	0	0

TABLE 10.4

Total Relation Matrix and D + R, D − R Values

	CR1	CR2	CR3	CR4	CR5	CR6	CR7	CR8	CR9	D + R	D − R
CR1	0.0592	0.1201	0.1032	0.0865	0.2422	0.1862	0.1537	0	0.1567	2,1217	0,0939
CR2	0.1290	0.0564	0.1711	0.0895	0.2522	0.1602	0.1555	0	0.1619	2,064	0,2876
CR3	0.0856	0.0803	0.0589	0.0801	0.2245	0.1404	0.1069	0	0.1429	2,1173	−0,2781
CR4	0.1181	0.0483	0.1247	0.0458	0.2268	0.1095	0.1051	0	0.1432	1,8052	0,0378
CR5	0.1367	0.1281	0.1490	0.1278	0.1385	0.1996	0.1968	0	0.2296	3,2754	−0,6632
CR6	0.0756	0.0708	0.0825	0.0707	0.1487	0.0628	0.1597	0	0.1300	2,1931	−0,5915
CR7	0.0282	0.0265	0.0328	0.0264	0.1134	0.1061	0.0409	0	0.1063	1,8123	−0,8511
CR8	0.1987	0.1863	0.2479	0.1859	0.3229	0.2048	0.1967	0	0.2404	1,7836	1,7836
CR9	0.1828	0.1714	0.2276	0.1710	0.3001	0.2227	0.2164	0	0.1280	3,059	0,181

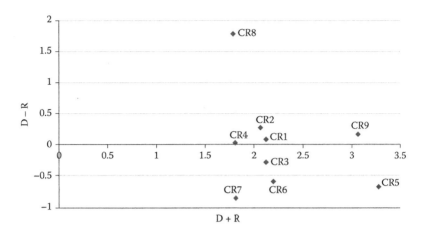

FIGURE 10.1
Cause and effect diagram.

According to Table 10.4 and Figure 10.1, the evaluation criteria, namely, green purchasing (CR1), green design (CR2), green information systems (CR4), regulations (CR8), and pressures from stakeholders (CR9) are classified in the cause group. Product recovery (CR3), environmental performance (CR5), economic performance (CR6), and operational performance (CR7) are included in the effect group.

10.2 Other Models

Fuzzy DEMATEL was used by Lin (2013) to analyze the interrelationships among the factors (green supply-chain management practices, organizational performance, and external driving factors) affecting the green supply-chain management implementation.

10.3 Conclusions

In this chapter, a model was presented illustrating the use of DEMATEL to solve an ECMPRO-related problem. An overview of another model was also presented in the chapter.

Reference

Lin R-J. Using fuzzy DEMATEL to evaluate the green supply chain management practices. *Journal of Cleaner Production* 2013;40: 32–9.

11

TOPSIS

The technique for order of preference by similarity to ideal solution (TOPSIS) is frequently used to solve various environmentally conscious manufacturing and product recovery (ECMPRO) problems, including the evaluation of recycling, remanufacturing, reverse logistics initiatives, and ranking of suppliers based on environmental factors. The high level of uncertainty associated with those problems forces researchers to integrate fuzzy logic into TOPSIS. Hence, nearly all of the proposed methodologies are based on fuzzy TOPSIS. In this chapter, we present two TOPSIS-based models. The first model uses TOPSIS to evaluate two recycling programs offered by different companies. The second one ranks the alternative recycling partners of a company.

11.1 The First Model (Evaluation of Recycling Programs)

In this section, we present a fuzzy TOPSIS model to evaluate recycling programs offered by two different companies.

11.1.1 Success Factors for a Recycling Program

Self-explanatory success factors for the implementation of a recycling program can be listed as follows:

1. Having enough knowledge about the success factors for the implementation of the recycling program (KNO)
2. Customer awareness about the recycling program (CAW)
3. Simplicity of the recycling program (SRP)
4. Convenient disposal of used products at collection centers (CON)
5. Incentives for disposal of used products (INC)
6. Effectiveness of collection methods (ECM)
7. Provision of necessary information about used products being collected (PNI)
8. Regular collection of used products (RCP)

9. Use of special mechanisms for abusers of the recycling program (SMA)

10. Selecting good locations for the centers where reprocessed goods are sold (GLA)

11. Provision of incentives to buyers of reprocessed goods (PIN)

12. Recycling program owners' cooperation with local government (COP)

11.1.2 Ranking of Recycling Programs Using Fuzzy TOPSIS

Public opinion is very important for the success of a recycling program. That is why, in this section, the opinions of three representatives of a community are used to determine importance levels of the success factors listed in Section 11.1.1. Table 11.1 presents the linguistic weights given by them to each success factor. These linguistic weights are converted into triangular fuzzy numbers (TFNs) using Table 11.2. Another TFN, called the *average weight*, is then formed by averaging those TFNs (see Table 11.3). The normalized

TABLE 11.1

Linguistic Weights of Success Factors

Factor	Rep. 1	Rep. 2	Rep. 3
KNO	Medium	High	Low
CAW	High	Very high	Very low
SRP	Low	High	Medium
CON	High	Medium	Low
INC	Very high	Very high	Medium
ECM	Medium	Medium	High
PNI	High	High	High
RCP	Very high	High	Very high
SMA	Medium	High	Medium
GLA	Very high	Very high	High
PIN	High	Very high	High
COP	Medium	High	Very high

TABLE 11.2

Conversion Table for Weights of Success Factors

Linguistic Weight	TFN
Very high	(0.75, 1.00, 1.00)
High	(0.50, 0.75, 1.00)
Medium	(0.25, 0.50, 0.75)
Low	(0.00, 0.25, 0.50)
Very low	(0.00, 0.00, 0.25)

TABLE 11.3

Average Weights of Success Factors

Factor	Average Weight
KNO	(0.25, 0.50, 0.75)
CAW	(0.42, 0.58, 0.75)
SRP	(0.25, 0.50, 0.75)
CON	(0.25, 0.50, 0.75)
INC	(0.58, 0.83, 0.92)
ECM	(0.33, 0.58, 0.83)
PNI	(0.50, 0.75, 1.00)
RCP	(0.67, 0.92, 1.00)
SMA	(0.33, 0.58, 0.83)
GLA	(0.67, 0.92, 1.00)
PIN	(0.58, 0.83, 1.00)
COP	(0.50, 0.75, 0.92)
Sum	(5.33, 8.24, 10.5)

weight of each success factor is then calculated by taking the ratio of the average weight of each success factor to the sum of the average weights of all the success factors (see Table 11.4).

Assuming that there are two recycling programs to be conducted by two different companies, these programs are linguistically rated by the three representatives considering each success factor. Then, the linguistic ratings are converted into TFNs using Table 11.5. The resulting decision matrix is presented in Table 11.6 (RP1 and RP2 are the recycling programs).

TABLE 11.4

Normalized Weights of Success Factors

Factor	Average Weight
KNO	(0.02, 0.06, 0.14)
CAW	(0.04, 0.07, 0.14)
SRP	(0.02, 0.06, 0.14)
CON	(0.02, 0.06, 0.14)
INC	(0.06, 0.10, 0.17)
ECM	(0.03, 0.07, 0.16)
PNI	(0.05, 0.09, 0.19)
RCP	(0.06, 0.11, 0.19)
SMA	(0.03, 0.07, 0.16)
GLA	(0.06, 0.11, 0.19)
PIN	(0.06, 0.10, 0.19)
COP	(0.05, 0.09, 0.17)

TABLE 11.5

Triangular Fuzzy Numbers
Associated with Linguistic Ratings

Linguistic Rating	TFN
Very good	(8, 10, 10)
Good	(5, 8, 10)
Fair	(4, 6, 8)
Poor	(0, 4, 6)
Very poor	(0, 0, 4)

TABLE 11.6

Decision Matrix

Factor	RP1	RP2
KNO	(8, 10, 10)	(0, 0, 4)
CAW	(4, 6, 8)	(8, 10, 10)
SRP	(4, 6, 8)	(4, 6, 8)
CON	(0, 0, 4)	(8, 10, 10)
INC	(8, 10, 10)	(5, 8, 10)
ECM	(5, 8, 10)	(4, 6, 8)
PNI	(0, 4, 6)	(4, 6, 8)
RCP	(0, 0, 4)	(8, 10, 10)
SMA	(4, 6, 8)	(5, 8, 10)
GLA	(5, 8, 10)	(5, 8, 10)
PIN	(0, 4, 6)	(8, 10, 10)
COP	(0, 0, 4)	(5, 8, 10)

Having a completed decision matrix, the six stages of TOPSIS can then be implemented as follows:

Stage 1: *Formation of the normalized decision matrix.* The normalized decision matrix depicted in Table 11.7 is obtained by calculating r_{ij} for each element of the decision matrix depicted in Table 11.6.

Stage 2: *Formation of the weighted normalized decision matrix.* Table 11.8 presents the weighted normalized decision matrix, which is formed by multiplying the normalized weights of success factors given in Table 11.4 with the normalized decision matrix presented in Table 11.7.

Stage 3: *Calculation of ideal and negative-ideal solutions.* For each row in the weighted normalized decision matrix, a maximum TFN (ideal solution) and a minimum TFN (negative-ideal solution) is determined. In this study, it is assumed that the TFN with the highest most promising quantity (the second parameter in the TFN) is the

TABLE 11.7

Normalized Decision Matrix

Factor	RP1	RP2
KNO	(0.74, 1.00, 1.25)	(0.00, 0.00, 0.50)
CAW	(0.31, 0.52, 0.90)	(0.63, 0.86, 1.12)
SRP	(0.35, 0.71, 1.41)	(0.35, 0.71, 1.41)
CON	(0.00, 0.00, 0.50)	(0.74, 1.00, 1.25)
INC	(0.57, 0.78, 1.06)	(0.35, 0.62, 1.06)
ECM	(0.39, 0.80, 1.56)	(0.31, 0.60, 1.25)
PNI	(0.00, 0.56, 1.50)	(0.40, 0.83, 2.00)
RCP	(0.00, 0.00, 0.50)	(0.74, 1.00, 1.25)
SMA	(0.31, 0.60, 1.25)	(0.39, 0.80, 1.56)
GLA	(0.35, 0.71, 1.41)	(0.35, 0.71, 1.41)
PIN	(0.00, 0.37, 0.75)	(0.69, 0.93, 1.25)
COP	(0.00, 0.00, 0.80)	(0.46, 1.00, 2.00)

TABLE 11.8

Weighted Normalized Decision Matrix

Factor	RP1	RP2
KNO	(0.02, 0.06, 0.18)	(0.00, 0.00, 0.07)
CAW	(0.01, 0.04, 0.13)	(0.03, 0.06, 0.16)
SRP	(0.01, 0.04, 0.20)	(0.01, 0.04, 0.20)
CON	(0.00, 0.00, 0.07)	(0.02, 0.06, 0.18)
INC	(0.03, 0.08, 0.18)	(0.02, 0.06, 0.18)
ECM	(0.01, 0.06, 0.25)	(0.01, 0.04, 0.20)
PNI	(0.00, 0.05, 0.29)	(0.02, 0.08, 0.38)
RCP	(0.00, 0.00, 0.10)	(0.04, 0.11, 0.24)
SMA	(0.01, 0.04, 0.20)	(0.01, 0.06, 0.25)
GLA	(0.02, 0.08, 0.27)	(0.02, 0.08, 0.27)
PIN	(0.00, 0.04, 0.14)	(0.04, 0.09, 0.24)
COP	(0.00, 0.00, 0.14)	(0.02, 0.09, 0.34)

maximum and the TFN with the lowest most promising quantity is the minimum. Ideal and negative-ideal solutions are presented in Table 11.9. Defuzzified values are given in Table 11.10.

Stage 4: *Calculation of the separation distances.* The separation distance for each recycling program is then calculated (see Table 11.11).

Stage 5: *Calculation of the relative closeness coefficient.* Next, the relative closeness coefficient is calculated for each recycling program (see Table 11.12).

Stage 6: *Ranking of the preference order.* RP2 is better than RP1, since its closeness coefficient (0.70) is higher than that of RP1 (0.30).

TABLE 11.9

Ideal and Negative-Ideal Solutions

Factor	Ideal Solution	Negative-Ideal Solution
KNO	(0.02, 0.06, 0.18)	(0.00, 0.00, 0.07)
CAW	(0.03, 0.06, 0.16)	(0.01, 0.04, 0.13)
SRP	(0.01, 0.04, 0.20)	(0.01, 0.04, 0.20)
CON	(0.02, 0.06, 0.18)	(0.00, 0.00, 0.07)
INC	(0.03, 0.08, 0.18)	(0.02, 0.06, 0.18)
ECM	(0.01, 0.06, 0.25)	(0.01, 0.04, 0.20)
PNI	(0.02, 0.08, 0.38)	(0.00, 0.05, 0.29)
RCP	(0.04, 0.11, 0.24)	(0.00, 0.00, 0.10)
SMA	(0.01, 0.06, 0.25)	(0.01, 0.04, 0.20)
GLA	(0.02, 0.08, 0.27)	(0.02, 0.08, 0.27)
PIN	(0.04, 0.09, 0.24)	(0.00, 0.04, 0.14)
COP	(0.02, 0.09, 0.34)	(0.00, 0.00, 0.14)

TABLE 11.10

Defuzzified Ideal and Negative-Ideal Solutions

Factor	Ideal Solution	Negative Ideal Solution
KNO	0.09	0.02
CAW	0.08	0.06
SRP	0.08	0.08
CON	0.09	0.02
INC	0.10	0.09
ECM	0.11	0.08
PNI	0.16	0.11
RCP	0.13	0.03
SMA	0.11	0.08
GLA	0.12	0.12
PIN	0.12	0.06
COP	0.15	0.05

TABLE 11.11

Separation Distances of Recycling Programs

RecyclingProgram	S^*	S^-
RP1	0.180	0.077
RP2	0.077	0.180

TABLE 11.12

Relative Closeness Coefficients of Recycling Programs

Recycling Program	C^*
RP1	0.30
RP2	0.70

11.2 The Second Model (Selection of Recycling Partners)

Increasing customer awareness toward environmental issues and stricter government regulations force many manufacturers to engage in recycling activities. That is why selection of recycling partners is a crucial topic for supply-chain managers (Wittstruck and Teuteberg 2012). In this section, we analyze the use of TOPSIS in the selection of recycling partners.

11.2.1 Determination of the Selection Criteria

The following criteria are used to evaluate alternative recycling partners:

- Financial capability (FC)
- IT capacity/interfaces (IT)
- Price (PR)
- Effective environmental management information system (ES)
- Quality of recycling processes (QR)
- Recycling capability (RC)
- Standardized health and safety conditions (HS)
- Sustainable image (SI)

11.2.2 Ranking Recycling Partners Using TOPSIS

Suppose that there are four candidate recycling partners (A, B, C, and D); each recycling partner is evaluated for each criterion. After converting the linguistic evaluations into numbers using Table 11.13, the decision matrix can be formed as given in Table 11.14.

Having a completed decision matrix, the six stages of TOPSIS can be implemented as follows:

Stage 1: *Formation of the normalized decision matrix.* The normalized decision matrix depicted in Table 11.15 is obtained by calculating r_{ij} for each element of the decision matrix depicted in Table 11.14.

TABLE 11.13

Linguistic Ratings and Values

Linguistic Rating	Value
Very bad	0.15
Bad	0.25
Satisfactory	0.50
Good	0.75
Very good	1.00

TABLE 11.14

Decision Matrix

	FC	IT	PR	ES	QR	RC	HS	SI
A	0.50	1	0.75	0.25	1	0.75	0.25	1
B	0.75	0.25	1	1	0.25	0.75	0.50	0.25
C	1	0.25	0.50	0.75	0.75	1	0.50	0.25
D	0.75	0.50	0.50	0.25	1	0.25	1	0.75

TABLE 11.15

Normalized Decision Matrix

	FC	IT	PR	ES	QR	RC	HS	SI
A	0.32	0.85	0.52	0.19	0.62	0.51	0.20	0.77
B	0.49	0.21	0.70	0.77	0.15	0.51	0.40	0.19
C	0.65	0.21	0.35	0.58	0.46	0.68	0.40	0.19
D	0.49	0.43	0.35	0.19	0.62	0.17	0.80	0.58

Stage 2: *Formation of the weighted normalized decision matrix.* Table 11.16 presents the weighted normalized decision matrix, which is formed by multiplying the normalized decision matrix presented in Table 11.15 by the normalized weights of the criteria. In this example, the normalized weights of the criteria are taken as (0.08, 0.08, 0.15, 0.15, 0.17, 0.13, 0.12, 0.12).

Stage 3: *Calculation of ideal and negative-ideal solutions.* Each column in the weighted normalized decision matrix shown in Table 11.16 has a minimum rank and a maximum rank. They are the ideal and the negative-ideal solutions, respectively, for the corresponding criterion. For example (see Table 11.16), with respect to criterion IT, the ideal solution (maximum rank) is 0.07, and the negative-ideal solution (minimum rank) is 0.02.

Stage 4: *Calculation of the separation distances.* The separation distance for each recycling program is then calculated (see Table 11.17).

TABLE 11.16

Weighted Normalized Decision Matrix

	FC	IT	PR	ES	QR	RC	HS	SI
A	0.03	0.07	0.08	0.03	0.11	0.07	0.02	0.10
B	0.04	0.02	0.11	0.12	0.03	0.07	0.05	0.02
C	0.05	0.02	0.05	0.09	0.08	0.09	0.05	0.02
D	0.04	0.03	0.05	0.03	0.11	0.02	0.10	0.07

TABLE 11.17

Separation Distances

Recycling Partner	S^*	S^-
A	0.13	0.14
B	0.14	0.12
C	0.13	0.11
D	0.14	0.13

TABLE 11.18

Relative Closeness Coefficients

Recycling Partner	C^*
A	0.52
B	0.46
C	0.46
D	0.48

Stage 5: *Calculation of the relative closeness coefficient.* Next, the relative closeness coefficient is calculated for each recycling program (see Table 11.18).

Stage 6: *Ranking of the preference order.* Since the best alternative is the one with the highest relative closeness coefficient, the preference order for the recycling partners is A, D, (B or C) (which means that A is the best recycling partner).

11.3 Other Models

Researchers have developed many TOPSIS-based methodologies for the evaluation of various remanufacturing, recycling, or reverse logistics initiatives. The option selection problem in reverse logistics was solved by Wadhwa et al. (2009) using fuzzy TOPSIS. Remery et al. (2012) developed a methodology for end-of-life (EOL) option selection using TOPSIS. Gao et al. (2010) evaluated green design options using fuzzy TOPSIS. Yeh and Xu (2013) developed a fuzzy TOPSIS approach to evaluate the alternative recycling activities of a company, based on a number of environmental, economic, and social criteria. Five sustainable concepts (design for environment, life cycle assessments, environmental conscious quality function deployment, theory of inventive problem solving, and life cycle impact assessment) were evaluated in Vinodh et al. (2013) by using fuzzy TOPSIS. Mahapatara et al. (2013) evaluated various reverse manufacturing alternatives (remanufacturing, reselling, repairing, cannibalization, and refurbishing) by employing fuzzy

TOPSIS. Green Supply Chain Management (GSCM) practices of an automotive company were assessed in Diabat et al. (2013) based on the results of a fuzzy TOPSIS study.

TOPSIS was also used to evaluate third-party reverse logistic providers and environmental performance of suppliers. Interpretive structural modeling and fuzzy TOPSIS were integrated in Kannan et al. (2009) to evaluate the alternative third-party reverse logistics providers. Fuzzy TOPSIS models for the environmental performance evaluation of suppliers were presented in Awasthi et al. (2010), Govindan et al. (2012), and Shen et al. (2013).

11.4 Conclusions

In this chapter, two models were presented for the use of TOPSIS to solve ECMPRO-related problems. In the first model, two recycling programs offered by different companies were evaluated, while the second model ranked alternative recycling partners of a company. An overview of other models was also presented in the chapter.

References

Awasthi A, Chauhan SS, Goyal SK. A fuzzy multicriteria approach for evaluating environmental performance of suppliers. *International Journal of Production Economics* 2010;126: 370–8.

Diabat A, Khodaverdi R, Olfat L. An exploration of green supply chain practices and performances in an automotive industry. *The International Journal of Advanced Manufacturing Technology* 2013;68: 949–61.

Gao Y, Liu Z, Hu D, Zhang L, Gu G. Selection of green product design scheme based on multi-attribute decision-making method. *International Journal of Sustainable Engineering* 2010;3: 277–91.

Govindan K, Khodaverdi R, Jafarian A. A fuzzy multi criteria approach for measuring sustainability performance of a supplier based on triple bottom line approach. *Journal of Cleaner Production* 2012;47: 345–54.

Kannan G, Pokharel S, Sasi Kumar P. A hybrid approach using ISM and fuzzy TOPSIS for the selection of reverse logistics provider. *Resources, Conservation and Recycling* 2009;54: 28–36.

Mahapatara SS, Sharma SK, Parappagoudar MB. A novel multi-criteria decision making approach for selection of reverse manufacturing alternative. *International Journal of Services and Operations Management* 2013;15: 176–95.

Remery M, Mascle C, Agard B. A new method for evaluating the best product end-of-life strategy during the early design phase. *Journal of Engineering Design* 2012;23: 419–41.

Shen L, Olfat L, Govindan K, Khodaverdi R, Diabat A. A fuzzy multi criteria approach for evaluating green supplier's performance in green supply chain with linguistic preferences. *Resources, Conservation and Recycling* 2013;74: 170–9.

Vinodh S, Mulanjur G, Thiagarajan, A. Sustainable concept selection using modified fuzzy TOPSIS: A case study. *International Journal of Sustainable Engineering* 2013;6: 109–16.

Wadhwa S, Madaan J, Chan FTS. Flexible decision modeling of reverse logistics system: A value adding MCDM approach for alternative selection. *Robotics and Computer-Integrated Manufacturing* 2009;25: 460–9.

Wittstruck D, Teuteberg F. Integrating the concept of sustainability into the partner selection process: A fuzzy-AHP-TOPSIS approach. *International Journal of Logistics Systems and Management* 2012;12: 195–226.

Yeh C-H, Xu Y. Sustainable planning of e-waste recycling activities using fuzzy multicriteria decision making. *Journal of Cleaner Production* 2013:52: 194–204.

12

ELECTRE

Elimination and choice translating reality (ELECTRE) is an outranking methodology. The use of ELECTRE for the solution of environmentally conscious manufacturing and product recovery (ECMPRO)-related problems is very limited. In this chapter, we present one ELECTRE-based methodology that evaluates alternative environmentally conscious manufacturing programs (Rao 2009).

12.1 The Model

The following success factors are considered for the evaluation of environmentally conscious manufacturing programs:

- Cost (CST)
- Percentage of recyclable material (REC)
- Percentage reduction in process waste (PRC)
- Percentage reduction in packaging waste (PAC)
- Compliance with government regulations (GOV)

The analytical hierarchy process (AHP) can be used to determine the importance weights of the criteria. Comparative importance values of the criteria are presented in Table 12.1. The normalized eigenvector of the comparative importance values matrix is also presented in this table. This vector represents the relative weights given by the decision maker to the criteria (i.e., $w_{CST} = 0.340$, $w_{REC} = 0.094$, $w_{PRC} = 0.322$, $w_{PAC} = 0.144$, $w_{GOV} = 0.100$).

Next, the decision matrix is completed as shown in Table 12.2.

The steps of ELECTRE can then be implemented as follows:

Step 1: The decision matrix is normalized. Each element of the decision matrix is normalized as depicted in Table 12.3. For instance, the

TABLE 12.1

Comparative Importance Values of Criteria

	CST	REC	PRC	PAC	GOV	Normalized Eigenvector
CST	1	3	1	3	4	0.340
REC	1/3	1	1/2	1/2	1/2	0.094
PRC	1	2	1	3	4	0.322
PAC	1/3	2	1/3	1	2	0.144
GOV	1/4	2	1/4	1/2	1	0.100

TABLE 12.2

Decision Matrix

	CST	REC	PRC	PAC	GOV
ECMP1	600,000	15	7	4	40
ECMP2	300,000	12	5	10	35
ECMP3	400,000	25	10	12	42
ECMP4	200,000	30	3	5	32
ECMP5	500,000	28	8	7	25

TABLE 12.3

Normalized Decision Matrix

	CST	REC	PRC	PAC	GOV
ECMP1	0.632456	0.289858	0.445399	0.218870	0.506451
ECMP2	0.316228	0.231887	0.318142	0.547176	0.443144
ECMP3	0.421637	0.483097	0.636285	0.656611	0.531773
ECMP4	0.210819	0.579717	0.190885	0.273588	0.405161
ECMP5	0.527046	0.541069	0.509028	0.383023	0.316532

normalized rating of ECMP3 with respect to success factor REC can be calculated as follows:

$$r_{32} = \frac{25}{\sqrt{15^2 + 12^2 + 25^2 + 30^2 + 28^2}} = 0.483097$$

Step 2: The weighted normalized decision matrix $(V = (v_{ij})_m \times {}_n)$ *is constructed.* Table 12.4 presents the weighted normalized decision matrix, which is formed by multiplying the weights of success factors with the normalized decision matrix presented in Table 12.3. For instance, the weighted normalized rating of ECMP4 with respect to success factor PRC can be calculated as follows:

$$v_{43} = 0.322 \cdot 0.190885 = 0.061465$$

TABLE 12.4

Weighted Normalized Decision Matrix

	CST	REC	PRC	PAC	GOV
ECMP1	0.215035	0.027247	0.143419	0.031517	0.050645
ECMP2	0.107517	0.021797	0.102442	0.078793	0.044314
ECMP3	0.143357	0.045411	0.204884	0.094552	0.053177
ECMP4	0.071678	0.054493	0.061465	0.039397	0.040516
ECMP5	0.179196	0.050860	0.163907	0.055155	0.031653

Step 3: The concordance and discordance sets are determined.

$C_{12} = \{2,3,5\}$	$C_{21} = \{1,4\}$	$C_{31} = \{1,2,3,4,5\}$	$C_{41} = \{1,2,4\}$	$C_{51} = \{1,2,3,4\}$
$C_{13} = \{\}$	$C_{23} = \{1\}$	$C_{32} = \{2,3,4,5\}$	$C_{42} = \{1,2\}$	$C_{52} = \{2,3\}$
$C_{14} = \{3,5\}$	$C_{24} = \{3,4,5\}$	$C_{34} = \{3,4,5\}$	$C_{43} = \{1,2\}$	$C_{53} = \{2\}$
$C_{15} = \{5\}$	$C_{25} = \{1,4,5\}$	$C_{35} = \{1,3,4,5\}$	$C_{45} = \{1,2,5\}$	$C_{54} = \{3,4\}$
$D_{12} = \{1,4\}$	$D_{21} = \{2,3,5\}$	$D_{31} = \{\}$	$D_{41} = \{3,5\}$	$D_{51} = \{5\}$
$D_{13} = \{1,2,3,4,5\}$	$D_{23} = \{2,3,4,5\}$	$D_{32} = \{1\}$	$D_{42} = \{3,4,5\}$	$D_{52} = \{1,4,5\}$
$D_{14} = \{1,2,4\}$	$D_{24} = \{1,2\}$	$D_{34} = \{1,2\}$	$D_{43} = \{3,4,5\}$	$D_{53} = \{1,3,4,5\}$
$D_{15} = \{1,2,3,4\}$	$D_{25} = \{2,3\}$	$D_{35} = \{2\}$	$D_{45} = \{3,4\}$	$D_{54} = \{1,2,5\}$

Step 4: Concordance matrix (C) is calculated. For instance the concordance index of $c(1,2)$ is calculated as follows:

$$c(1,2) = w_2 + w_3 + w_5 = 0.094 + 0.322 + 0.1 = 0.516$$

Thus,

$$C = \begin{bmatrix} - & 0.516 & 0 & 0.422 & 0.1 \\ 0.484 & - & 0.434 & 0.566 & 0.584 \\ 1 & 0.66 & - & 0.566 & 0.906 \\ 0.578 & 0.434 & 0.434 & - & 0.534 \\ 0.9 & 0.416 & 0.094 & 0.466 & - \end{bmatrix}$$

Step 5: Disconcordance matrix (D) is calculated. For instance, the discon-cordance index of $d(1,2)$ is calculated as follows:

$$d(1,2) = \frac{\max\big(|0.215035 - 0.107517|, |0.031517 - 0.078793|\big)}{\max\big(|0.215035 - 0.107517|, |0.027247 - 0.021797|, |0.102442 - 0.143419|, |0.031517 - 0.078793|, |0.050645 - 0.044314|\big)}$$

$$d(1,2) = 1$$

Thus,

$$D = \begin{bmatrix} - & 1 & 1 & 1 & 1 \\ 0.051 & - & 1 & 0.875 & 0.858 \\ 0 & 0.35 & - & 0.5 & 0.133 \\ 0.572 & 1 & 1 & - & 0.953 \\ 0.530 & 1 & 1 & 1 & - \end{bmatrix}$$

Step 6: Concordance dominance matrix (E) is obtained by comparing each element of concordance matrix against average index of concordance ($\bar{c}$):

$$\bar{c} = \frac{(0.516 + 0 + \ldots + 0.466)}{5(5-1)} = 0.5047$$

Thus,

$$E = \begin{bmatrix} - & 1 & 0 & 0 & 0 \\ 0 & - & 0 & 1 & 1 \\ 1 & 1 & - & 1 & 1 \\ 1 & 0 & 0 & - & 1 \\ 1 & 0 & 0 & 0 & - \end{bmatrix}$$

Step 7: Disconcordance dominance matrix (F) is obtained by comparing each element of disconcordance matrix against average index of disconcordance ($\bar{d}$):

$$\bar{d} = \frac{1 + 1 + \ldots + 1}{5(5-1)} = 0.7411$$

Thus,

$$F = \begin{bmatrix} - & 1 & 1 & 1 & 1 \\ 0 & - & 1 & 1 & 1 \\ 0 & 0 & - & 0 & 0 \\ 0 & 1 & 1 & - & 1 \\ 0 & 1 & 1 & 1 & - \end{bmatrix}$$

Step 8: The net superior (c_j) and inferior values (d_j) are calculated. For instance,

TABLE 12.5

Ranking of Environmentally Conscious Manufacturing Programs

	Net Superior Values	Net Inferior Values	Ranking of Net Superior Values	Ranking of Net Inferior Values
ECMP1	−1.924	2.847	5	5
ECMP2	−0.052	0.566	2	3
ECMP3	2.264	−3.017	1	1
ECMP4	−0.076	0.15	3	2
ECMP5	−0.248	0.586	4	4

$$c_1 = (0.516 + 0 + 0.422 + 0.1) - (0.484 + 1 + 0.578 + 0.9) = -1.924$$

$$d_1 = (1 + 1 + 1 + 1) - (0.051 + 0 + 0.572 + 0.530) = 2.847$$

The rest of the values are presented in Table 12.5.

According to Table 12.5, ECMP3 is ranked first considering both the net superior and inferior values.

12.2 Other Models

Bufardi et al. (2004) used ELECTRE III to evaluate five end-of-life (EOL) alternatives by considering four criteria, including EOL treatment cost, human health, ecosystem quality, and resources.

12.3 Conclusions

In this chapter, a model is presented for the use of ELECTRE to solve ECMPRO-related problems. This model evaluated the alternative environmentally conscious manufacturing programs. An overview of other models is also presented in the chapter.

References

Bufardi A, Gheorghe R, Kiritsis D, Xirouchakis P. Multicriteria decision-aid approach for product end-of-life alternative selection. *International Journal of Production Research* 2004;42: 3139–57.

Rao RV. An improved compromise ranking method for evaluation of environmentally conscious manufacturing programs. *International Journal of Production Research* 2009;47: 4399–4412.

13

PROMETHEE

Preference ranking organization methods for enrichment evaluations (PROMETHEE) (Brans 1985) is a multiple criteria decision making (MCDM) method allowing the partial (PROMETHEE I) as well as the complete ranking (PROMETHEE II) of alternatives. Evaluation of environmentally conscious manufacturing and product recovery (ECMPRO) programs, third-party reverse logistic providers, green suppliers, and product recovery options are the main ECMPRO issues for which researchers have developed PROMETHEE-based MCDM methodologies. In this chapter, we present a PROMETHEE-based model that uses PROMETHEE to balance a disassembly line.

13.1 The Model

The disassembly line balancing problem (DLBP) is an important and actively researched problem in ECMPRO (see McGovern and Gupta [2011] for more information on DLBP). It is a multiobjective problem as described by Gungor and Gupta (2002), and has been mathematically proven to be NP-complete by McGovern and Gupta (2007), which makes the desire to achieve the best balance computationally expensive when considering large-sized problems.

The application of the PROMETHEE-based disassembly line balancing approach (Avikal et al. 2013a,b, 2014) will be demonstrated using the product structure presented by Gungor and Gupta (2002). Table 13.1 presents the data of the model.

The following criteria were considered in the example:

- Disassembly time (DT): To maximize the utilization of workstations, the disassembly tasks with high disassembly times should be assigned to the workstations as early as possible.

- Demand (DM): Highly demanded parts should be disassembled as early as possible because the probability of damage will be higher as the time a part spends in the disassembly line increases.

- Hazardousness (HR): Any problem occurring during the disassembly of hazardous parts may result in the stoppage of the disassembly line or may damage the demanded parts. That is why hazardous parts must be removed as early as possible.

TABLE 13.1

Knowledge Base for the Example

Part Number	Part Description	Time	Hazardous	Demand	Predecessor(s)
1	PC top cover	14	No	360	–
2	Floppy drive	10	No	500	1
3	Hard drive	12	No	620	1
4	Back plane	18	No	480	7
5	PCI cards	23	No	540	1
6	RAM modules (2)	16	No	750	2 or 3
7	Power supply	20	Yes	295	8
8	Motherboard	36	No	720	2, 3, 5, 6

Source: Gungor, A. and Gupta, S.M., *International Journal of Production Research*, 40, 2569–2589, 2002.

The analytical hierarchy process (AHP) can be used to determine the weights of the disassembly line balancing criteria. Comparative importance values presented in Table 13.2 were formed based on the input from decision makers. The normalized eigenvector associated with the comparison matrix was also presented in this table. According to this vector, the weights of the disassembly line balancing criteria were determined as follows: $w_{DT} = 0.11$, $w_{DM} = 0.26$, $w_{HR} = 0.63$.

Following the determination of criteria and criteria weights, preference functions associated with the criteria should be determined (Tables 13.3 and 13.4).

TABLE 13.2

Comparative Importance Values of Criteria

	DT	DM	HR	Normalized Eigenvector
DT	1	1/3	1/5	0.11
DM	3	1	1/3	0.26
HR	5	3	1	0.63

TABLE 13.3

Difference Values for "Disassembly Time" Criterion

$d(DT_i, DT_j)$	DT1	DT2	DT3	DT4	DT5	DT6	DT7	DT8
DT1	0	4	2	−4	−9	−2	−6	−22
DT2	−4	0	−2	−8	−13	−6	−10	−26
DT3	−2	2	0	−6	−11	−4	−8	−24
DT4	4	8	6	0	−5	2	−2	−18
DT5	9	13	11	5	0	7	3	−13
DT6	2	6	4	−2	−7	0	−4	−20
DT7	6	10	8	2	−3	4	0	−16
DT8	22	26	24	18	13	20	16	0

TABLE 13.4

Joint Preference Functions for "Disassembly Time" Criterion

$P(DT_i, DT_j)$	DT1	DT2	DT3	DT4	DT5	DT6	DT7	DT8
DT1	0	0.4	0.2	0	0	0	0	0
DT2	0	0	0	0	0	0	0	0
DT3	0	0.2	0	0	0	0	0	0
DT4	0.4	0.8	0.6	0	0	0.2	0	0
DT5	0.9	1	1	0.5	0	0.7	0.3	0
DT6	0.2	0.6	0.4	0	0	0	0	0
DT7	0.6	1	0.8	0.2	0	0.4	0	0
DT8	1	1	1	1	1	1	1	0

Preference indices for all disassembly task pairs are calculated as presented in Table 13.5. For instance, the preference index for the DT1 and DT3 pair is calculated as follows:

$$\pi(DT1, DT3) = 0.11 \cdot 0.2 + 0.26 \cdot 0 + 0.63 \cdot 0 = 0.022$$

Table 13.6 presents the leaving (Φ^+) and entering flows (Φ^-) of disassembly tasks. For instance, the Φ^+ and Φ^- values for Disassembly task 2 are calculated as follows:

$$\phi^+ = \frac{0.441 + 0.063 + 0.63}{8 - 1} = 0.16$$

$$\phi^- = \frac{0.044 + 0.4 + 0.088 + 0.236 + 0.696 + 0.37 + 0.74}{8 - 1} = 0.37$$

TABLE 13.5

Preference Indices

$P(DT_i, DT_j)$	DT1	DT2	DT3	DT4	DT5	DT6	DT7	DT8
DT1	0	0.044	0.022	0	0	0	0.20475	0
DT2	0.441	0	0	0.063	0	0	0.63	0
DT3	0.63	0.4	0	0.441	0.252	0	0.63	0
DT4	0.422	0.088	0.066	0	0	0.022	0.58275	0
DT5	0.666	0.236	0.11	0.244	0	0.077	0.663	0
DT6	0.652	0.696	0.4535	0.63	0.63	0	0.63	0.0945
DT7	0.326	0.37	0.348	0.282	0.26	0.304	0	0.26
DT8	0.74	0.74	0.425	0.74	0.677	0.11	0.74	0

TABLE 13.6

Leaving and Entering Flows of Disassembly Tasks

Disassembly Task	Φ^+	Φ^-
1	0.04	0.55
2	0.16	0.37
3	0.34	0.20
4	0.17	0.34
5	0.29	0.26
6	0.54	0.07
7	0.31	0.58
8	0.60	0.05

Net flows and ranks of disassembly tasks are presented in Table 13.7. For instance, the net flow for Disassembly task 2 is calculated as follows:

$$\phi^{net} = 0.16 - 0.37 = -0.21$$

Assuming a cycle time of 40 s for the disassembly line, the assignment of disassembly tasks to the workstations is carried out as follows. Although Disassembly task 1 has the lowest rank in Table 13.7, it has to be assigned to the first workstation according to the precedence relationships presented in Table 13.1. After the assignment of Disassembly task 1, there are three eligible disassembly tasks: 2, 3, and 5. Since Disassembly task 3 has the highest rank, it becomes the second disassembly task assigned to Workstation 1. After this assignment, there are three disassembly tasks that can be assigned to Workstation 1: 2, 5, and 6. Although Disassembly task 6 has the highest rank, its disassembly time (16 s) exceeds the remaining time in Workstation 1 (40 − (14 + 12) = 14 s). Disassembly task 5, with a disassembly time of 23 s, cannot

TABLE 13.7

Net Flows and Ranks of Disassembly Tasks

Disassembly Task	Φ^{net}	Rank
1	−0.51	8
2	−0.21	6
3	0.14	3
4	−0.17	5
5	0.03	4
6	0.47	2
7	−0.27	7
8	0.55	1

TABLE 13.8

Assignment of Disassembly Tasks to Workstations

Station	Disassembly Task	Disassembly Time	Station Time	Idle Time
1	1	14		
	3	12		
	2	10	36	4
2	6	16		
	5	23	39	1
3	8	36	36	4
4	7	20		
	4	18	38	2

be assigned to Workstation 1 for the same reason. Hence, Disassembly task 2 becomes the third disassembly task assigned to Workstation 1. The assignment of three disassembly tasks to Workstation 1 results in an idle time of 4 min [40 – (14 + 10 + 12)]. The assignment of the remaining disassembly tasks to the other workstations is carried out in a similar fashion. Table 13.8 presents the workstation assignments for all disassembly tasks.

13.2 Other Models

Avikal et al. (2013b) used PROMETHEE to balance a disassembly line. Avikal et al. (2013a) integrated AHP and PROMETHEE to develop a disassembly line balancing methodology. First, the weight of each criterion was calculated using AHP. Then, PROMETHEE ranked the disassembly tasks to be assigned to the disassembly line stations. Avikal et al. (2014) modified this approach by determining criteria weights with fuzzy AHP instead of AHP.

Alternative fasteners in design for disassembly were evaluated in Ghazilla et al. (2013) by using PROMETHEE. Tuzkaya et al. (2009) integrated fuzzy ANP and fuzzy PROMETHEE to evaluate the environmental performance of suppliers.

13.3 Conclusions

In this chapter, a model was presented for the use of PROMETHEE to solve an ECMPRO-related problem. An overview of other models was also presented in the chapter.

References

Avikal S, Mishra PK, Jain R. An AHP and PROMETHEE methods-based environment friendly heuristic for disassembly line balancing problems. *Interdisciplinary Environmental Review* 2013a;14: 69–85.

Avikal S, Mishra PK, Jain R. A Fuzzy AHP and PROMETHEE method-based heuristic for disassembly line balancing problems. *International Journal of Production Research* 2014;52: 1306–17.

Avikal S, Mishra PK, Jain R, Yadav HC. A PROMETHEE method based heuristic for disassembly line balancing problem. *Industrial Engineering & Management Systems* 2013b;12: 254–63.

Brans JP, Vincke P. A preference ranking organization method: The PROMETHEE method for MCDM. *Management Science* 1985;31: 647–656.

Ghazilla R, Taha Z, Yusoff S, Rashid S, Sakundarini N. Development of decision support system for fastener selection in product recovery oriented design. *The International Journal of Advanced Manufacturing Technology* 2013;70: 1403–13.

Gungor A, Gupta SM. Disassembly line in product recovery. *International Journal of Production Research* 2002;40: 2569–89.

McGovern SM, Gupta SM. A balancing method and genetic algorithm for disassembly line balancing. *European Journal of Operational Research* 2007;179: 692–708.

McGovern SM, Gupta SM. *The Disassembly Line: Balancing and Modeling*. McGraw-Hill, New York, 2011.

Tuzkaya G, Ozgen A, Ozgen D, Tuzkaya UR. Environmental performance evaluation of suppliers: A hybrid fuzzy multi criteria decision approach. *International Journal of Environment Science and Technology* 2009;6, 477–90.

14

VIKOR

VlseKriterijumska Optimizacija I Kompromisno Resenje (VIKOR) is a multiple criteria decision making (MCDM) methodology developed especially for the solution of multicriteria problems with noncommensurable and conflicting criteria. It provides a *compromise solution* which guarantees a maximum *group utility* for the *majority* and a minimum of an individual regret for the *opponent*. Since environmentally conscious manufacturing and product recovery (ECMPRO) problems involve many conflicting criteria, VIKOR can be an effective solution methodology. However, the number of studies using VIKOR to solve ECMPRO problems is very limited. In this chapter, we present a VIKOR-based model to evaluate alternative reverse logistics operating channels. It uses the analytical hierarchy process (AHP) and VIKOR to evaluate alternative reverse logistics operating channels.

14.1 The Model

Self-explanatory success factors for the evaluation of alternative reverse logistics operating channels are listed here:

- Flexibility (FL)
- Successful setup and operation of information technology (IT) applications
- Response time (RT)
- Cost (CT)
- Management control over reverse logistics operations (MC)

Comparative importance values of the criteria are presented in Table 14.1. The normalized eigenvector of the comparative importance values matrix is also presented in this table. This vector represents the relative weights given by the decision maker to the criteria (i.e., $w_1 = 0.110$, $w_2 = 0.078$, $w_3 = 0.207$, $w_4 = 0.463$, $w_5 = 0.142$).

TABLE 14.1

Comparative Importance Values of Criteria

	FL	IT	RT	CT	MC	Normalized Eigenvector
FL	1	2	1/3	1/5	1	0.110
IT	1/2	1	1/2	1/5	1/2	0.078
RT	3	2	1	1/2	1	0.207
CT	5	5	2	1	4	0.463
MC	1	2	1	1/4	1	0.142

There are three most commonly used reverse logistics operating channels:

- *Manufacturer operation (MAN)*: The used products are collected by the manufacturer. All necessary resources (e.g., information systems, human resources, etc.) required for reverse logistics activities are managed by the manufacturer.
- *Joint operation (JON)*: Collection of the used products is carried out by the retailer. The other reverse logistics-related operations are managed by the manufacturer.
- *Third-party reverse logistics provider (3PRLP)*: Reverse logistics operations are outsourced to a third-party reverse logistics provider.

Each reverse logistics operating channel alternative is evaluated for each criterion using a 10-point scale. The resulting decision matrix is presented in Table 14.2.

Using the completed decision matrix, the five stages of VIKOR can be implemented as follows:

Step 1: *Determine the best f_i^* and the worst f_i^- values associated with each criterion.* According to the decision matrix presented in Table 14.2, the best f_i^* and the worst f_i^- values associated with each criterion are presented in Table 14.3.

Step 2: *Compute S and R values for each alternative.* Calculate S and R values for each alternative (see Table 14.4). For example, the S and R values associated with the second alternative (JON) are calculated as follows:

TABLE 14.2

Decision Matrix

	FL	IT	RT	CT	MC
MAN	4	5	4	3	10
JON	2	3	4	5	5
3PRLP	8	10	8	10	3

TABLE 14.3

The Best f_i^* and the Worst f_i^- Values

	FL	IT	RT	CT	MC
f_i^*	8	10	8	10	10
f_i^-	2	3	4	3	3

TABLE 14.4

S_j and R_j Values

	S_j	R_j
MAN	0.799	0.463
JON	0.827	0.331
3PRLP	0.142	0.142

$$S_2 = 0.110 \cdot \frac{(8-2)}{(8-2)} + 0.078 \cdot \frac{(10-3)}{(10-3)} + 0.207 \cdot \frac{(8-4)}{(8-4)}$$

$$+0.463 \cdot \frac{(10-5)}{(10-3)} + 0.142 \cdot \frac{(10-5)}{(10-3)} = 0.827$$

$$R_2 = \max\left[0.110 \cdot \frac{(8-2)}{(8-2)}, 0.078 \cdot \frac{(10-3)}{(10-3)}, 0.207 \cdot \frac{(8-4)}{(8-4)}, \right.$$

$$\left. 0.463 \cdot \frac{(10-5)}{(10-3)}, 0.142 \cdot \frac{(10-5)}{(10-3)} \right] = 0.331$$

Step 3: *Compute the Q value for each alternative.* Calculate the Q value associated with each alternative. Table 14.5 presents the Q values of reverse logistics operating channel alternatives for different v values. For instance, the Q value of the first alternative (MAN) for $v = 0.25$ is calculated as follows:

$$Q_1 = 0.25 \cdot \frac{(0.799 - 0.142)}{(0.827 - 0.142)} + (1 - 0.25) \cdot \frac{(0.463 - 0.142)}{(0.463 - 0.142)} = 0.990$$

TABLE 14.5

Q_j Values Calculated for Different v Values

	$Q_j(v = 0.00)$	$Q_j(v = 0.25)$	$Q_j(v = 0.50)$	$Q_j(v = 0.75)$	$Q_j(v = 1.00)$
MAN	1	0.990	0.980	0.969	0.959
JON	0.588	0.691	0.794	0.897	1
3PRLP	0	0	0	0	0

TABLE 14.6

Ranking of the Alternatives According to S, R, and Q Values

	S_j	R_j	Q_j $(v=0.00)$	Q_j $(v=0.25)$	Q_j $(v=0.50)$	Q_j $(v=0.75)$	Q_j $(v=1.00)$
MAN	2	3	3	3	3	3	2
JON	3	2	2	2	2	2	3
3PRLP	1	1	1	1	1	1	1

TABLE 14.7

Checking for Conditions 1 and 2 for Different v Values

	$v = 0.00$	$v = 0.25$	$v = 0.50$	$v = 0.75$	$v = 1.00$
$Q(a'')$	0.588	0.691	0.794	0.897	0.959
$Q(a')$	0	0	0	0	0
$Q(a'') - Q(a')$	0.588	0.691	0.794	0.897	0.959
DQ	0.5	0.5	0.5	0.5	0.5
Condition 1	Satisfied	Satisfied	Satisfied	Satisfied	Satisfied
Condition 2	Satisfied	Satisfied	Satisfied	Satisfied	Satisfied

Step 4: *Rank the alternatives, sorting by the values S, R, and Q, in descending order.* Final ranking of alternatives by S, R, and Q values is presented in Table 14.6.

Step 5: *Check for the conditions.* The two conditions of VIKOR are checked for different values of v (see Table 14.7). According to this table, Conditions 1 and 2 are satisfied for all values of v. That is why we can assert that the alternative with the smallest value of Q (3PRLP) has an acceptable advantage and stability over the two other reverse logistics operating channel alternatives.

14.2 Other Models

In Rao (2009), environmentally conscious manufacturing programs are evaluated by using VIKOR. Datta et al. (2012) integrate VIKOR and fuzzy set theory to select green suppliers. Samantra et al. (2013) evaluate the product recovery options by using the methodology proposed by Datta et al. (2012). The two-step methodology proposed by Sasikumar and Haq (2011) supports the design of a closed-loop supply chain. First, a suitable third-party reverse logistics provider (3PRLP) is selected using VIKOR. Then, various design decisions, including raw material procurement, production, and distribution are taken based on the results of a mixed-integer linear-programming model.

14.3 Conclusions

In this chapter, a model was presented for the use of VIKOR to solve ECMPRO-related problems in which alternative channels for reverse logistics operations were evaluated. An overview of other models was also presented in the chapter.

References

Datta S, Samantra C, Mahapatra SS, Banerjee S, Bandyopadhyay A. Green supplier evaluation and selection using VIKOR method embedded in fuzzy expert system with intervalvalued fuzzy numbers. *International Journal of Procurement Management* 2012;5: 647–78.

Rao RV. An improved compromise ranking method for evaluation of environmentally conscious manufacturing programs. *International Journal of Production Research* 2009;47: 4399–4412.

Samantra C, Sahu NK, Datta S, Mahapatra SS. Decision-making in selecting reverse logistics alternative using interval-valued fuzzy sets combined with VIKOR approach. *International Journal of Services and Operations Management* 2013;14: 175–96.

Sasikumar P, Haq AN. Integration of closed loop distribution supply chain network and 3PRLP selection for the case of battery recycling. *International Journal of Production Research* 2011;49: 3363–85.

15

MACBETH

Measuring attractiveness by a categorical based evaluation technique (MACBETH) is a multiple criteria decision making method similar to the analytical hierarchy process (AHP). The main difference is that MACBETH uses an interval scale while AHP adopts a ratio scale. In this chapter, we present a MACBETH-based model to evaluate alternative third-party reverse logistics providers (3PRLPs).

15.1 The Model

An original equipment manufacturer (OEM) plans to evaluate five alternative 3PRLPs using the following criteria:

- Cooperation (COOP): The 3PRLP should be willing to work with the company. A close relationship must be formed between the 3PRLP and the company.
- Flexibility (FLEX): The 3PRLP should have the ability to satisfy the varying needs of the company regarding the timing, size, and location of orders.
- Advanced information system capabilities (INFO): The 3PRLP should provide advanced information system capabilities such as online tracking.
- Cost of reverse logistics operations (COST): All reverse logistics-related cost components (transportation cost, reprocessing cost, etc.) should be considered while evaluating a 3PRLP.

15.1.1 Evaluating 3PRLPs Using M-MACBETH Software

The MACBETH value tree presented in Figure 15.1 was constructed by entering the four evaluation criteria into M-MACBETH software. Cooperation, flexibility, and advanced information system capabilities are beneficial criteria, while the cost of reverse logistics operations is the nonbeneficial criterion.

FIGURE 15.1
MACBETH value tree.

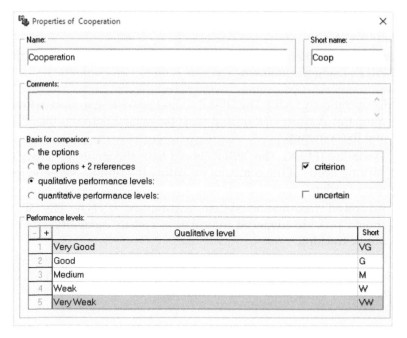

FIGURE 15.2
Performance levels for cooperation criterion.

A five-level qualitative scale was proposed for the beneficial criteria. The performance levels in this scale are arranged as "very good" (VG), "good" (G), "medium" (M), "weak" (W), and "very weak" (VW). For an example, see Figure 15.2 for the performance levels for the cooperation criterion.

The qualitative scale defined for the nonbeneficial criterion (cost of reverse logistics operations) also involves five levels, "very low" (VL), "low" (L), "medium" (M), "high" (H), and "very high" (VH). Thus, the performance levels for the cost of reverse logistics operations criterion are given in Figure 15.3.

The difference of attractiveness between performance levels are evaluated for each criterion. For instance, pairwise comparison matrices of performance levels for the flexibility and cost of reverse logistics operations criteria are given in Figures 15.4 and 15.5, respectively.

Next, the pairwise comparison matrix for the criteria is constructed and criteria weights are determined as shown in Figure 15.6.

The table of performances is created by evaluating the performance of each alternative 3PRLP for each criterion as shown in Figure 15.7.

The M-MACBETH software then determines the overall scores for the five 3PRLP alternatives as shown in Figure 15.8.

FIGURE 15.3
Performance levels for cost of reverse logistics operations criterion.

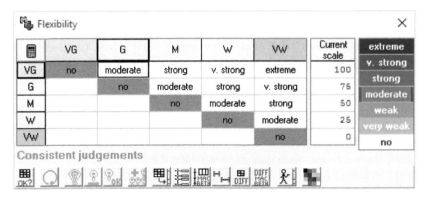

FIGURE 15.4
Pairwise comparison of performance levels for flexibility criterion.

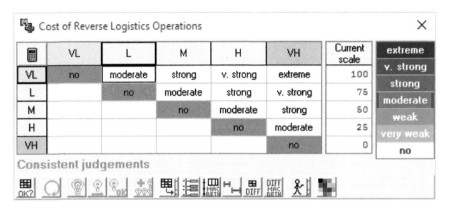

FIGURE 15.5
Pairwise comparison of performance levels for cost of reverse logistics operations criterion.

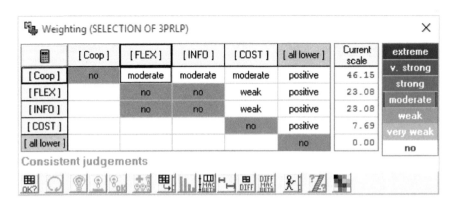

FIGURE 15.6
Determination of criteria weights.

Table of performances ✕

Options	Coop	FLEX	INFO	COST
3PRLP-1	VG	M	M	M
3PRLP-2	G	G	M	H
3PRLP-3	VG	W	G	H
3PRLP-4	M	M	G	M
3PRLP-5	M	G	G	VH

FIGURE 15.7
Table of performances.

Table of scores ✕

Options	Overall	Coop	FLEX	INFO	COST
3PRLP-1	73.08	100.00	50.00	50.00	50.00
3PRLP-2	65.39	75.00	75.00	50.00	25.00
3PRLP-3	71.15	100.00	25.00	75.00	25.00
3PRLP-4	55.77	50.00	50.00	75.00	50.00
3PRLP-5	57.70	50.00	75.00	75.00	0.00
[all upper]	100.00	100.00	100.00	100.00	100.00
[all lower]	0.00	0.00	0.00	0.00	0.00
Scaling constants:		0.4615	0.2308	0.2308	0.0769

FIGURE 15.8
Table of scores.

15.2 Other Model

Dhouib (2014) proposes a fuzzy MACBETH methodology to evaluate options in reverse logistics for waste automobile tires.

15.3 Conclusions

In this chapter, a model was presented for the use of MACBETH to solve an environmentally conscious manufacturing and product recovery (ECMPRO)-related problem. An overview of another model was also presented in the chapter.

Reference

Dhouib D. An extension of MACBETH method for a fuzzy environment to analyze alternatives in reverse logistics for automobile tire wastes. *Omega* 2014;42: 25–32.

16

Gray Relational Analysis

Gray relational analysis (GRA) is a very effective strategy when there is incomplete information regarding one or more parameters of a decision problem. Due to uncertainties associated with the quantities and conditions of end-of-life (EOL) products, GRA is a promising solution methodology for environmentally conscious manufacturing and product recovery (ECMPRO) problems. However, the use of GRA in ECMPRO problems is limited to a few studies. In this chapter, we present a GRA-based model for the evaluation of suppliers by considering environmental as well as traditional performance criteria.

16.1 The Model

The following environmental criteria were considered:

- Carbon footprint (CFP)
- Environment-friendly packaging (EFP)

There are four performance evaluation criteria:

- Price (PRC)
- Delivery (DEL)
- Quality (QLY)
- Cooperation (CPR)

Prior to the application of the GRA procedure, criteria weights are determined using the analytical hierarchy process (AHP). The pairwise comparison matrix is presented in Table 16.1.

The company evaluates three potential suppliers for a component. Table 16.2 presents the evaluation results of the suppliers.

The decision matrix presented here was constructed based on Table 16.2:

$$
\begin{bmatrix}
120 & 5 & 75 & 4 & 5 & 3 \\
90 & 4 & 100 & 3 & 4 & 4 \\
100 & 3 & 85 & 4 & 4 & 5
\end{bmatrix}
$$

TABLE 16.1

Comparative Importance Values of Criteria

	CFP	EFP	PRC	DEL	QLY	CPR	Norm. Eigenvector
CFP	1	4	1/4	1/2	1/5	1/2	0.074
EFP	1/4	1	1/7	1/5	1/8	1/4	0.030
PRC	4	7	1	3	1/2	4	0.287
DEL	2	5	1/3	1	1/3	1	0.121
QLY	5	8	2	3	1	4	0.381
CPR	2	4	1/4	1	1/4	1	0.107

TABLE 16.2

Evaluation of Alternative Suppliers

	CFP	EFP	PRC	DEL	QLY	CPR
Supplier 1	120	5	75	4	5	3
Supplier 2	90	4	100	3	4	4
Supplier 3	100	3	85	4	4	5

Step 1: Determination of referential and comparison series: The referential series (x_0) were determined as follows:

$$x_0 = (90, 5, 75, 4, 5, 5)$$

Comparison series were determined as follows:

$$x_1 = (120, 5, 75, 4, 5, 3)$$

$$x_2 = (90, 4, 100, 3, 4, 4)$$

$$x_3 = (100, 3, 85, 4, 4, 5)$$

Step 2: Construction of normalized decision matrix: Depending on the type of the criterion (larger-is-better or smaller-is-better), two normalization approaches were applied. For instance, the normalized value of carbon footprint criterion for Supplier 1 was calculated as follows:

$$x_1^*(1) = \frac{120 - 120}{120 - 90} = 0$$

The normalized value of the quality criterion for Supplier 1 was calculated as follows:

$$x_1^*(5) = \frac{5-4}{5-4} = 1$$

Step 3: Calculation of the gray relational coefficient: Gray relational coefficient values were then calculated (Table 16.3). The distinguished coefficient was taken as 0.5. For instance, the gray relational coefficient value of Supplier 1 for the carbon footprint criterion was calculated as follows:

$$r_1(1) = \frac{0+0.5*1}{1+0.5*1} = 0.33$$

Step 4: Calculation of the degree of gray equation coefficient: The degree of gray equation coefficient for each supplier is calculated as follows:

$$\tau(\text{Supplier 1}) = 0.33*0.074+1*0.30+1*0.287+1*0.121$$

$$+1*0.381+0.33*0.107 = 0.879$$

$$\tau(\text{Supplier 2}) = 1*0.074+0.5*0.30+0.33*0.287+0.33*0.121$$

$$+0.33*0.381+0.5*0.107 = 0.406$$

$$\tau(\text{Supplier 3}) = 0.33*0.074+0.33*0.30+0.56*0.287$$

$$+1*0.121+0.33*0.381+1*0.107 = 0.549$$

According to gray relation grades, the best supplier is Supplier 1. The alternative suppliers can be ranked as follows:

Supplier 1 > Supplier 3 > Supplier 2

TABLE 16.3

Gray Relational Values of Alternative Suppliers

	CFP	EFP	PRC	DEL	QLY	CPR
Supplier 1	0.33	1	1	1	1	0.33
Supplier 2	1	0.5	0.33	0.33	0.33	0.5
Supplier 3	0.33	0.33	0.56	1	0.33	1

16.2 Other Models

GRA is used in Chan (2008) to evaluate EOL processing alternatives considering uncertainty and several criteria at the material level. Li and Zhao (2009) developed a green supplier selection methodology by integrating the threshold method and GRA. Chen et al. (2010) integrated fuzzy logic and GRA to evaluate suppliers based on green criteria. Dou et al. (2014) integrated the analytical network process and GRA for the valuation of green supplier development programs.

16.3 Conclusions

In this chapter, a model was presented for the use of GRA to solve an ECMPRO-related problem. An overview of other models was also presented in the chapter.

References

Chan JWK. Product end-of-life options selection: grey relational analysis approach. *International Journal of Production Research* 2008;46: 2889–912.

Chen CC, Tseng ML, Lin YH, Lin ZS. Implementation of green supply chain management in uncertainty, in: *IEEE International Conference on Industrial Engineering and Engineering Management*, 2010, pp. 260–4.

Dou Y, Zhu Q, Sarkis J. Evaluating green supplier development programs with a grey-analytical network process-based methodology. *European Journal of Operational Research* 2014;233: 420–31.

Li X, Zhao C. Selection of suppliers of vehicle components based on green supply chain, in: *16th International Conference on Industrial Engineering and Engineering Management*, Beijing, 2009, pp. 1588–91.

17

Conclusions

Due to stricter environmental regulations and decreasing natural resources, environmentally conscious manufacturing and product recovery (ECMPRO) initiatives have received increased attention from researchers and practitioners in recent years. Although there are various algorithms, models, heuristics, and software that have been developed to solve ECMPRO-related problems, multiple criteria decision making (MCDM) techniques have lately received increasing attention from scholars due to their ability to consider multiple and conflicting objectives simultaneously. Application of MCDM techniques to ECMPRO-related problems is a new and fast-growing area of interest. While there are many books on the market that emphasize the application of MCDM techniques to various problems such as sustainable energy and transportation, land-use management, and selection of engineering materials in product design, as of now, none exclusively focuses on the issues that arise in the area of ECMPRO.

This book addressed the issues that arise in the application of MCDM techniques to the problems associated with ECMPRO. The MCDM techniques considered in this book included goal programming, fuzzy goal programming, linear physical programming, data envelopment analysis, analytical hierarchy process, analytical network process, DEMATEL, TOPSIS, ELECTRE, PROMETHEE, VIKOR, MACBETH, and gray relational analysis. The issues addressed in this book may serve as foundations to build bodies of knowledge by other scholars in this new and fast-growing field of research.

Subject Index

Author Index